THE SECOND
NATURE
OF THINGS

WILL CURTIS

THE SECOND

NATURE

OF THINGS

With Wood Engravings by Nora S. Unwin

THE ECCO PRESS

The Ecco Press
100 West Broad Street
Hopewell, NJ 08525
Published simultaneously in Canada by
Penguin Books Canada Ltd., Ontario
Printed in the United States of America
Designed by Frank Lieberman
First Paperback Edition

Permissions and copyright notices for all contributed material are
found on pages 281–290.

The publisher and author are grateful to the Sharon Arts Center, Sharon,
New Hampshire, for the use of wood engravings by Nora S. Unwin in this book.

Library of Congress Cataloging-in-Publication Data

Curtis, Will, 1917–
    The second nature of things / Will Curtis; illustrations by Nora
S. Unwin.—1st ed.
      p.   cm.
    1. Natural history.   I. Title.
QH81.C876   1992
508—dc20   92-16806
ISBN 0-88001-285-4
ISBN 0-88001-383-4 (paperback)

The text of this book is set in Meridien with headings in Optima.

9 8 7 6 5 4 3 2

TO MY MOTHER AND FATHER,
DOROTHY MERCER CURTIS
JOHN ARNOLD CURTIS

Contents

Preface

It was on our dairy farm where my real interest in nature began. Much of my time was spent in the fields plowing, planting, haying, and, in the early spring, in the sugarbush gathering sap buckets. Now and then, I would stop, look, and listen. An amazing series of natural events evolved around me. My evenings were often spent reading magazines and books on nature, trying to learn about what I had seen in the fields and woods during the day.

After our herd of Jersey cattle had been sold, Jane and I found ourselves the owners of a bookstore. The object of a bookstore is, of course, to sell books. To help in this we went on the air with reviews on local radio stations.

Now established as something of a radio personality, I was asked by the Vermont Institute of Natural Science, a non-profit institution known for its wildlife educational programs, to do a radio program for them.

That was some fourteen years ago. The program turned into an approximately three-minute spot named "The Nature of Things," aided and promoted by Vermont Public Radio.

I am grateful to National Public Radio for giving me a slot in the weekend edition of "All Things Considered." Eventually, as a result of a conversation I had with Les Line, then editor of *Audubon* magazine, Vermont Public Radio became the recipient of yearly grants for four fruitful years from The National Audubon Society.

When The National Audubon Society turned their plans from radio to television, Vermont Public Radio found a grant for the continuance of "The Nature of Things" from Waste Management Inc., a national company offering comprehensive treatment and disposal options of hazardous waste. This has assured the continued growth of the program on stations around the country.

Researching the material for a five-day-a-week nationally syndicated program is constant, but the education that comes with it is rewarding. The more I learn about nature the more there is to know. This means that I must subscribe to some forty-five magazines and publications to keep me abreast of the changing scene in all phases of the natural world. Whether we are traveling or at home, my wife Jane is a constant help with editing and writing.

Letters from listeners in various parts of the country bring me the satisfaction that I am in some way adding interest to their lives, and the fact that The Ecco Press asked me to do a second volume of *The Nature of Things* is very gratifying indeed. *The Second Nature of Things* includes over two hundred of my more recent nationally aired programs, and focuses on the ecological problems plaguing our planet today.

My programming plans for the future will continue to be related to wildlife and its myriad world, with views of the planetary systems, the oceans, archaeology, people, plants, animals, and the soil. All these and many more constitute the nature of things.

—Will Curtis
Hartland, Vermont

nature (nā'tyủr, -chủr)[F., from L. *nātūra*, from *nāt*-, p.p. stem of *nascī* to be born], *n.* The essential qualities of anything; the physical or psychical constitution of a person or animal; natural character or disposition; kind, sort, class; the inherent energy or impulse determining these; vital or animal force; the whole sum of things, forces, activities, and laws constituting the physical universe; the physical power that produces the phenomena of the material world; this personified; the sum of physical things and forces regarded as distinct from man; the material universe regarded as distinct from the supernatural or from a creator; the natural condition of man preceding social organization; the undomesticated condition of animals or plants; unregenerate condition as opp. to a state of grace; nakedness; fidelity to nature in art.

THE SEASONS

SPRING WALK

It is a blowy, blustery March Sunday. A redwing blackbird trills from the top of the maple, calling us to get out and greet the spring. Put on your boots, explore the back roads, we hear him say, leave winter tasks behind in the house. The road we choose we often use during the summer and think we know it well. But how much we have missed! How could we not have seen that giant, an ancient maple. We try putting our arms around its massive, seamed trunk but can't even see the other's outstretched hand.

Spring is the time to see what lies beyond the road's edges before the trees take on their canopy of leaves. In summer you might as well be traveling in a tunnel, a pleasant green one, but a tunnel for all you can see.

We poke around ancient cellar holes, some abandoned so long ago that full-grown trees stand in the depressions. One cellar hole is still new. Two hundred years ago the settler had carefully laid the foundation of his house with mammoth granite blocks. He wanted a house that would last his family for generations. Last winter a new owner, unused to the ways of wood-burning stoves, burned the old house down.

At noon we eat our sandwiches in a sunny corner of a half-forgotten cemetery. In their sunken graves lie those who had cleared the land and built the houses. Scraping moss off the tilted stones we read their names, Jerusha, Tabitha, Elisha, Isaiah, names that are out of fashion now. Would they be disheartened to see their houses gone, their hard-won fields grown up to woods again?

One hundred and fifty years ago their flocks of Merino sheep grazed on the then-open hills. Now, as we walk along the road, only a few of their pastures remain clear of trees. Nature has had its way with them.

Back in our farmhouse we feel that we have not only had a spring walk but a walk through history.　　　[1]

THE MAD MARCH HARE

The giant rabbit that brings eggs at Easter is clearly a pagan symbol of fertility. Dress him up in a silk waistcoat with a watch chain in an attempt to turn him into a civilized being, and he remains still the epitome of procreation. How does he compare with the real rabbit? Varying hares in peak years are prodigious generation begetters, with as many as 3,400 hares recorded from a single square mile in Ontario.

How does our long-eared progenitor accomplish this? Simple, according to American naturalists. In the first warm days of March he goes mad, the essence of his courtship activity. We have all heard of the mad March hare, baring his teeth at rival male hares, rising up on his long hind feet and thrashing out with his short front ones, hissing, swearing unrepeatable words beneath his breath. But this madness, reputedly, is not all challenge and attack against other courting males. A mad hare in spring has been known to carry on quite foolishly before the she-hare, sitting for hours under a fir bough as if in a swoon and then suddenly dashing off in frantic leaping circles. She-hares are seldom impressed by these antics or by the male hare's large brown moony eyes.

The Chinese, it is said, record the jack rabbit, or jack hare, as living on the moon—or at least as coming from the moon. Moon-struck the he-hare is when the pull of the spring tides of March reaches him.

Not so his British cousin. Despite what many eminent authorities in the past have said (Lewis Carroll, for one, pointed out to Alice that "in that direction lives a Hatter, and in that direction . . . lives a March Hare. . . . They're both mad!"). Recent research in Great Britain contests this characterization of the hare as turning loony in the spring. Two scientists from England named Greenwood and Holley, according to *Geo* magazine, say that after studying over two thousand photographs of males and females, very few boxing matches between hares were recorded at all—only seventeen—and, of these, only 2.3 happened in March. However, since these few winter boxings occurred always after dark (in January it is dark in England by 3:00 P.M.) and the summer ones all occurred in tall grass, making adequate photography doubtful, critics of the British research have questioned their conclusions.

People have reported mad March hares for centuries. Certainly no twosome of scientists named Greenwood and Holley are going to undo the wisdom of the ages with a fistful of photos of a few rabbits, who, we may remember, are known to be averse to appearing, wittingly or unwittingly, before the camera's eye, always preferring to drop suddenly down a hole, disappearing from sight.   [2]

CALVIN COOLIDGE'S GLORIOUS FOURTH

Best of all days, the one that caused the most excitement, was the Glorious Fourth. For months boys had been hoarding money, getting enough pennies, nickles, and dimes together to send out an order to Ohio to the Brazil Fireworks Company. $1.64 would buy enough four-inch salutes and Chinese firecrackers to last all day if a fellow was careful and didn't fire everything off at once. Small rockets and bombs (tissue-wrapped balls with caps mixed with pebbles) made a most satisfactory noise. For days afterward small boys claimed they couldn't hear their mothers calling, their ears rang so. But this was small beer compared to the annual Plymouth cannon contest.

The cannon was the result of a surge of patriotic fervor and paid for by public subscription of Plymouth folk and cast at Isaac Tyson's foundry. But which village should have possession? Over the years the boys of the Union and the Notch had vied for the honor of firing on the Fourth, until the issue was settled once and for all in 1892. No one is exactly sure after all these years just who was in the little raiding party from the Notch that crept down in the very early hours of the morning before the Fourth, but it is thought that Calvin Coolidge was on hand. Quietly they drew the five-hundred-pound cannon from its place at the Union and dragged it up the steep hill to the Notch. Some mastermind decided to hide it behind the pile of manure in the basement of the Wilder barn.

Early on the morning of the Fourth, it was dragged up behind the big doors. When the doors were open, floor planks were lifted up, rolling out the cannon without any of the crew being seen. A long lanyard was pulled, the cannon announced the dawn of the glorious day with a full-throated roar that rattled every window in the village. Before the bewildered sleepers could gather their wits, the cannon was quickly drawn inside with a long rope and the barn doors closed. Swiftly it was swabbed out, reloaded, and rolled out again to proclaim that the Republic still stood.

A lookout cannily placed at the top of the Notch announced the charge of the angry Union boys. They had heard the distant thunder, discovered their prize missing, and came running up the road to be met with John Wilder. "I'll fight the first of you and then everyone in turn!" Faced with such fierceness, the Union forces prudently retreated, and the cannon has remained at the Notch ever since. That afternoon Calvin Coolidge, by now a young Amherst student, gave a stirring speech on the glories of independence with every appearance of good behavior. [3]

THE ICEBOX

In our family we bewilder young relatives by referring to the "icebox." "What's an icebox?" they say. And we have to explain that we mean the refrigerator. Iceboxes were those big oak chests that used to stand on back porches near the kitchen in which were stored blocks of ice to keep the food cool. The iceman would come around once a week and deftly chip out a piece of ice that would just fit, but the householder had to remember to empty the drip pan every day or the floor would be flooded. Iceboxes went out in the thirties sometime, but on Monhegan Island, off the Maine coast, we used one until about 1965. I must say the refrigerator is a big improvement, even if we can't remember to call it by its correct name.

What did people do before iceboxes were invented? They used a springhouse if they were lucky enough to have one, or else they ate food that was probably spoiled or went without. It was Frederick Tudor of Boston who thought up the idea of having ice to use in the warm weather. In the 1840s or so he shipped a load of ice to Cuba and made a fortune, even though half of it had thawed by the time it got to Havana. That started everyone in the north thinking about ice, and soon, beside just about every pond and lake arose icehouses, big structures insulated with sawdust between their double walls. The ice started forming after Christmas and when it was twelve and fourteen inches thick, men ventured out on the glassy surface with saws, tongs, ice hooks, axes, and chisels.

I can remember watching the men drawing lines on the ice when I skated on the village pond, and then a horse-drawn plow followed the line, gouging it deeper. With an especially long handsaw, men would finish the cut and soon there would be big, white squares bobbing in the channels of black water. I can tell you we were careful not to play shinny near those channels! The cakes were pushed over to a ramp where they would be drawn by block and tackle up and into the icehouse. Stacked carefully, with room for air circulation, each row of ice was covered with sawdust. Cubes of ice are handy, but I think I miss those days when the iceman came and gave me a piece of blue ice to suck on a hot August day! [4]

HAYING

The skilled scytheman, like the skilled axman, was among the aristocrats of New England laborers, and was in high demand in his season. At haying time an army of haymakers, armed with scythe and snath and stone, walked down out of the thin-soiled hills to be employed from farm to farm in the fertile lowlands. The gangs of mowers took to the fields in the dew-heavy gray of early dawn, marching together, swinging together, whetting together, and—so the old saying went—urinating together.

The rapid adoption after the Civil War of the mowing machine, usually drawn by horses, wrought revolutionary changes in small ways and large. No longer did the farmer's wife have to board a tableful of scythemen. Whereas the scythers set out when the grass was wrapped with dew, the mowing machine did not go to work until late in the day, when all was dry. Generational disputes arose, with some old men maintaining that the machine ruined the goodness of the hay. Young men rode the machines, while old men hand mowed along the fences and around the boulders.

The generations also differed as to the degree to which hay had to be made, or dried, before it could safely be stowed in the mow. The old men warned against musty hay, while the young men complained that after the hay had been raked and then turned and cocked and capped and shaken out and turned and cocked and shaken out again it was all but worn out. And even the young men did not commence to mow until the grass's peak of goodness was well past.

The vast economic and urban expansion after the Civil

War created a large market for hay—especially timothy hay—in the cities, as the demands of short-haul transport were borne, often piteously, by the horse. In a large city thousands of horses were required simply to power the streetcars.

The mowing machine allowed the farmer to expand his farm or, at least, his haying operations, producing a surplus of hay which he could sell for cash. Of the New England states, Maine was noted for its hay exports, primarily to the Boston market. Come fall in a major exporting region like the Kennebec Valley, the dealers' hay-pressing crews, with their horse-powered hay presses, traveled from farm to farm. Hay was shipped by rail, by steamer, and by schooner, the latter carrying deck cargoes piled so high that the sails had to be deep-reefed and the weather chosen with very great care. "Prime Down East hay" was so advertised and it is claimed that Kennebec Valley hay was specified for Boston's pampered fire horses. [5]

FALL

As I walk through my woods the wild woodlands rustle, crunch, and tick.

There is an occasional churr from the red squirrels, a more frequent bark of the gray squirrels, and a rather common caw from the crows.

But the busy commotion of spring, and even the less frantic bustle of late summer largely have disappeared.

It is a time when frost stiffens the dying weeds and causes them to crunch rather than to yield silently underfoot. When the sun heats the morning air enough to triumph over frost, the tree leaves begin to tick, tick, tick as they lose their grip on twigs and descend from branch to branch, creating a tick sound each time they touch branch or firm leaf.

Once on the woodland path, the yellow, red, or brown leaves rustle as one walks through the woods slightly scattering these color components. To some extent, the side effects of a stroll cause the path and its borders to change their color pattern, like a gently tapped kaleidoscope.

The blue jays have become largely silent—indeed, they have become scarce. Every autumn there is a migration of blue jays both from and through my farm. The remaining blue jays seldom break the silence, unless they discover an owl or sharp-shinned hawk.

The timely birds that fit a woodland fall are modest in both color and voice. The drab sparrows blend with woodland browns. Their vocal exchanges barely surpass a chip sound, usually as subdued in volume as their plumage is in color.

The flowers of fall generally are less flashy than the open field flowers of late summer. The fall flowers may be ray-and-petal flowers closely related to late-summer blossoms. The asters, for instance. Or, they may be those rare, exquisite gentians which hold a small corner of the flower field guide for themselves, quite apart from the run-of-the-mill floral crowd.

Persons to whom the woods are unfamiliar suppose that they see a dying year in the autumn woods. But the woods do not share their secrets with just everyone. They seem particularly to have deceived poets. For autumn in the woods is not a death stage but change in life-style. It is a change less profound than the sleep which we mammals accept familiarly and therefore without alarm.

The emerging naked limbs may be stark to those not in the know. But they are just a yearly exercise in woodland draftsmanship, a sort of black, geometric architecture puncturing the often deep blue sky. [6]

FALL COLOR—
NO MYSTERY AT ALL

While raking the leaves on my lawn, I thought that one of the most important things to realize about fall color is that the color of leaves throughout the year is created by chemical pigments found in the leaf. During the growing season, while the leaves are healthy, the leaves appear green because of the predominance of a green pigment called chlorophyll. Chlorophyll is an important chemical in the process of photosynthesis, the process by which plants manufacture their own food. As fall approaches, deciduous trees slow their food production. The green pigment within the leaves (chlorophyll) is no longer utilized, and the chlorophyll is broken down into other chemicals. When the green chlorophyll degrades, other pigments are exposed, revealing a variety of possible colors. Although most leaves appear green throughout the growing season, there are also other color pigments contained within them. As the green pigments break down in the fall, the other color pigments are unmasked.

Different trees contain different masked color pigments, causing the spectrum of colors each fall. Because all individuals of a particular species contain the same type of leaf pigment, the autumn leaf color of any tree is fairly predictable. In other words, all sugar maples will appear mostly red; while all hickories will be mostly yellow each autumn.

The type of leaf pigment contained within leaves will determine their autumn color. Carotene, for instance, will produce yellow or orange leaves. (This pigment, by the way, is the same one that gives carrots their distinctive color.) Tannins (present in oak and beech tree leaves) produce a brown fall leaf hue. Anthocyanins produce blue, red, and purple shades.

As many people realize, the quality of fall color can vary a great deal from year to year. There are several reasons for that; but the most important factor is anthocyanin, or red pigment, production. Anthocyanin production is tied to sugar accumulation in the leaf. Low temperatures and bright sunlight lead to sugar accumulation and, thus, anthocyanin production. Warm sunny days, followed by cool nights, lead to lots of red leaves and the best fall color. Why, even oak tree leaves will become reddish brown in years when there is suitable weather.

Differences in individual leaf color and quality may also result from differences in light and temperature. Leaves in the canopy or at the edge of the woods, for example, receive more light and tend to have higher anthocyanin production. Leaves in the understory or in the middle of the woods receive less sunlight and are, therefore, more apt to be orange or yellow.

The mystery of autumn leaf color isn't a mystery at all; it's simply an effective mechanism for winter survival.   [7]

NUTS!

In fall the nuts are ripe and there is a scramble to store or eat them.

Actually, a nut's most important job is to feed a young seedling plant—a little nut tree—in the first stages of its life.

But wild animals, and humans, too, long ago discovered many nuts are delicious; so of all the nuts that grow from spring blossoms only a very few ever get a chance to grow seedlings.

Some of the more plentiful nut trees in ruffed grouse range are hickories, black walnuts, butternuts, beechnuts, and hazelnuts.

Black walnuts come in very tough shells. But squirrels are able to gnaw open the shells to get at the tasty kernels. Sometimes squirrels find someone else harvesting the nuts along with them. Youngsters are famous walnut gatherers.

Butternuts are very tasty themselves. But one of the

differences between man and wild animals is that man finds other things to do with something most creatures just consider a food. Butternuts are a good example. Certain Indians used oil from butternuts on their hair. Some soldiers in the Civil War made a dye from butternuts for their uniforms. Because of the color of their uniforms, the soldiers got the name "butternuts."

There are different kinds of hickory trees, and nuts from some of them taste as if they're made out of soap. Other kinds, however, are very good. Indians used to pound up the delicious ones to make flour for cakes.

A very highly prized nut is the hazelnut. People who seek hazelnuts have a lot of competition. They are also favorites of mice, deer, chipmunks, squirrels, and grouse, to name a few of the wild things after hazelnuts.

A lot of wild animals, including grouse, also love beechnuts. They are tiny little nuts that grow into great, gray trees. If enough beechnuts could be gathered they could be ground into and used as a flour. Back when people really had to use the wild foods they found to survive they would place sheets of cloth on the ground under a beechnut tree and let the seeds fall on the cloth. It was an easy way to gather the little seeds. [8]

CHRISTMAS EVE

At midnight on Christmas Eve, all the animals can talk. Probably everyone has heard this story. However, the story goes on to warn that anyone who hides behind the hay in the barn in order to hear them will not live out the night or the year. This is just one instance of the many tales of supernatural things which happen on Christmas Eve which are not for human eyes to witness and of the dire consequences that occur to the curious.

On Christmas Eve in Scandinavia the dead return to revisit their old homes. Everything is made ready for them—the fire lit, candles left burning, a feast of holiday food spread on the special cloth covering the table, including a jug of Yule ale. But no one dares to peek through the cracks in the door to see them, for surely if they did, they would join the dead when they returned to their graves before dawn.

People still believe in the magic powers of animals—not only that they gain the ability to speak but that cattle in the barns kneel in their stalls at the supposed hour of the birth of the Christ child. In fact, Howison in his *Sketches of Upper Canada* (written over a hundred years ago) passes on that an Indian told him that "on Christmas night all deer kneel and look up to the Great Spirit." Streams are believed to turn to blood, and water in wells to turn to wine. No one dares to test this, however, without regretting the act.

How different these earlier beliefs are from the more general ones held today about the appearance of Kris Kringle or Santa Claus, who always eats the cookies and drinks the milk or ale left for him to refresh himself after he has finished the work of delivering presents and trimming the tree.

In Scandinavia trolls make merry on Christmas Eve, celebrating with dancing and revelry. Their music, singing, and cavorting can be heard in the hills and from under their stones, and it is especially prudent to stay safely at home between cock-crowing and daybreak, when, according to ancient authority, it is "very dangerous to be abroad."

Even Shakespeare recognized the special character of this holy time:

> *The bird of dawning singeth all night long,*
> *And then, they say, no spirit dare stir abroad. . . .*
> *No fairy takes, nor witch hath power to charm,*
> *So hallowed and so gracious is the time.* [9]

THE CHRISTMAS TREE

The American Christmas tree, although firmly established now, is a relatively recent development. The first mention of Christmas trees in America is at the time of the American Revolution, but it wasn't the Americans who had them. The Hessian soldiers who were assisting King George III in his ill-fated war against the American colonists apparently set up Christmas trees to celebrate a German Christmas thousands of miles from their homes. As of 1776, the Germans were the only people who included a decorated evergreen tree in their Christmas celebrations.

If it weren't for the Germans—not only the Hessians, but the German immigrants who settled in America—chances are this country never would have adopted the tradition of the Christmas tree. The English who settled here were against Christmas celebrations altogether. The Puritans considered them pagan—and they were right. Many of our favorite Christmas rituals are derived from Teutonic and Roman customs more closely related to the winter solstice than to the birth of Christ.

In England in the early 1640s, Oliver Cromwell passed ordinances that forbade all Christmas festivities. December 25 was to be just another workday, with no feasting or "pagan" revelry. In 1659, the General Court of Massachusetts followed suit and made the observance of Christmas a penal offense. Eventually these laws were repealed, but they slowed down the evolution of Christmas customs in America.

The first official mention of a Christmas tree in America—besides the stories about the homesick Hessian soldiers—was not until 1804, when a Christmas tree was described as part of Christmas festivities in Fort Dearborn, Illinois. At this same time the Christmas tree was also beginning to spread from Germany to the rest of Europe. The early 1800s marked the appearance of Christmas trees in Finland, Denmark, Sweden, and Norway, as well as in Lancaster and

York, Pennsylvania, Philadelphia, Cincinnati, Richmond, Williamsburg, and Cleveland.

In 1841 the Christmas tree finally reached England. Queen Victoria's husband, Prince Albert, put one up in Windsor Castle. The English were neither immediate nor unanimous in their acceptance of the idea. No less an advocate of happy Christmases than Charles Dickens referred to the Christmas tree as "the new German toy."

By the 1850s Christmas trees were beginning to be big business in the United States. In 1851 a man named Mark Carr thought it might be lucrative to cut trees in the Catskills near where he lived and sell them in New York City. It was. By 1880, two hundred thousand trees from all over the Northeast were being sold in New York City.   [10]

WREATHS

In New Zealand on Christmas Day, families take a picnic lunch and go bask on one of the many beautiful golden sandy beaches. Because we here in the north have always equated winter and the holidays with snow, we forget that much of the rest of the world enjoys a warm, green Christmas. How important to us is our experience of the changing seasons! and of the contrasts these changes bring! When we make wreaths of greens to hang on our doors, do we not hope to charm spring to return? Yet we are not the first to decorate our houses with evergreens. Certainly the Romans and the Norse did for magical purposes. John Stow, writing in 1598 in his *Survey of London*, remarked that:

> *Against the time of Christmas, every man's house as also his parish churches, were decked with holly, ivy, bays, and whatsoever the season affordeth to be green.*

Vermont Folklife Center and the University of Vermont Center for Research on Vermont tell us the fruits of the red-berried elder last on the branches until January if grouse or red squirrels have not eaten them. High-bush cranberries usually make it to the table instead of into wreaths, but if the berries of the maple-leafed viburnum are found while still crimson—eventually they turn black—these are sometimes tucked into wreaths. Partridge or wintergreen trail along the ground, but the sharp eye easily discovers their bright red berries. Sometimes the red bunchberry is sought. Wherever the mountain ash or rowan tree escape from the homesteads of early settlers, especially the Scot who planted the ash tree in his dooryard to protect his house from harm, these clusters of bright berries enhance any wreath, provided the robins haven't gotten there first.

But why do we make our wreaths round and not some other shape? For practicality perhaps. Yet from time immemorial the circle has represented the sun, the life force. So when we make a circle of evergreen boughs, decorated with red berries, red the color of blood and essential to all human life and of fire and of the heat of passion, we follow a very old human custom. We draw on the natural world around us to express our belief that the round of life continues, and to charm or perhaps propitiate the elements and the season to return us to the other half of the year, to the verdant green and bright sun of summer. [11]

ST. STEPHEN'S DAY

According to the Christian calendar, the day after Christmas is dedicated to Saint Stephen. According to the carol, Good King Wenceslas looked out on the Feast of Stephen . . .

> *Then a poor man came in sight,*
> *Gathering winter fuel . . .*

Traditionally, St. Stephen's day was set aside to share with friends and neighbors. In England, the day is called

Boxing Day, not in a sporting sense, but because on this day the parish poor boxes, which had been filled on Christmas, are opened and the alms are given to the needy. On this day, too, public servants such as the postman, the local constable, the sexton, and the bell ringers visit households in the community and receive gifts for services rendered over the past year. Tavern keepers give back to their customers part of the price they paid for their meals.

Among the Italians who migrated to central Vermont in the late nineteenth century to work in the marble and granite quarries, December 26 or St. Stephen's Day was a time for neighborhood sharing. On the twenty-fifth, families and close friends had gathered for the Christmas feast and to exchange gifts. But on the twenty-sixth, casual friends and acquaintances are visited. At the turn of the century men customarily called from house to house in their neighborhoods, where they would be entertained with pieces of homemade cakes called *panettone* and shot glasses of *grappa*, a powerful homemade wine distillate. After exchanging pleasantries the visitors went on to call on the next neighbor. These ritual visitations began in mid-morning and continued the rest of the day. By the 1920s and '30s, the women wanted to go along, too, and enjoy sampling homemade wine, white grapes which had been soaked in *grappa*, and delicious *panettone*.

With the wives away from the house making visits with their husbands, who was left at home to give callers their cake and wine? No one. It became customary simply to leave the front door open, and for visitors to help themselves to the treats which were left on a tray for them. With no one to talk to, the callers spent their time looking over the gifts conspicuously placed around the tree.

But because it was more fun to visit a house full of people than to visit an empty house, the custom of calling on St. Stephen's Day began to fade away. By the 1950s fewer and fewer people were making the ritual house calls. Instead, people were drawn to after-Christmas sales and trading in unwanted gifts. Some old-timers, though, haven't given in to the new trend and still insist on visiting their neighbors on St. Stephen's Day.   [12]

THE MISTLETOE TRADITION

Mistletoe is a parasite that draws water and nutrients from its hardwood hosts. Its seeds find their way to potential hosts with the help of birds. In some cases the bird has a seed stuck to its beak and rubs the seed off on the bark of a tree it lands on. In other cases, the bird has eaten a mistletoe berry and deposits the undigested seed, along with its droppings, on a branch it happens to perch on.

When a mistletoe seed germinates, it sends a rootlike structure called a haustorium through the tree's bark into its conducting tissues. Through this haustorium the mistletoe absorbs the water and mineral salts it needs to survive. Its leaves contain chlorophyll, so the plant can manufacture its own food once the host tree has provided the necessary water and minerals.

The mistletoe of Christmas custom is actually a European species that doesn't grow in North America. One of our two common species is similar, so the distinction between the authentic European mistletoe and our own North American species doesn't matter. The custom of hanging mistletoe indoors at Christmastime is a widespread European custom, but the tradition of kissing a person who walks or stands under it is strictly English. The origin of the English practice is not clear, but it may have to do with several associations that primitive cultures had with the plant.

The ancient Celts, whose priests were called druids, considered mistletoe sacred, especially if it happened to be growing in an oak tree, which was quite rare in the parts of Europe and the British Isles where the Celts lived. If mistletoe was found in an oak tree, it became the occasion for a ritual harvest and feast. More ordinary mistletoe, simply because it stayed green all winter, was considered a place where woodland spirits might be hiding during the cold weather. It was brought indoors to offer the woodland spirits warmth and to decorate houses during the long dark days of winter.

Many groups besides the Celts also honored mistletoe. Interestingly enough, despite the fact that mistletoe is a parasite, it was considered a positive phenomenon in all the cultures that attached importance to it. It was used, for instance, in rituals having to do with peace, fertility, safety in battle, good luck, good health, and protection against evil spirits. It was also considered a curative for several diseases and an antidote for poisons.

One possible origin of the English tradition of kissing under the mistletoe might be the ancient Scandinavian belief that if enemies met under mistletoe, they had to lay down their weapons and declare a temporary truce. In more peaceful times, it might have occurred to someone familiar with the Scandinavian tradition to declare the presence of a mistletoe an excuse for a man to kiss a woman without the fear of reprisal. [13]

SNOW-SEASON SURVIVORS

When snow covers a forest, animals that don't hibernate or migrate to warmer areas must adapt to cold temperatures and the meager food available. For large animals such as moose and elk, winter is a season to be endured until warmer weather comes: although winter kills most green plants or buries them under deep snow, the animals must keep searching for food. Some small animals, however, use the snow cover to their advantage, becoming kings of the forest at least until spring.

Rodents such as voles dig intricate tunnels through the snow, where they find seeds and grasses, insulation from the cold "outside," and protection from predators such as red foxes and ermines. The snowpack that they dig through is made of hard and soft bands stacked like layers of cake. Some are thick and consistent, evidence of heavy snowfalls. Some are icy, signifying that they were once at the surface and have melted and refrozen. The lowest layers, made of a fragile network of ice crystals called "depth hoar," are the easiest to tunnel through. The middle layers are usually the heaviest: sometimes even oxygen cannot pass through them.

When deep, soft snow piles up, some large animals (especially elk and deer) sink and flounder, becoming easy prey for packs of wolves. Snowshoe hares avoid this particular problem by packing down lanes in the snow with their broad feet.

Icy crust, which forms when cold nights follow warm, sunny days, is also hazardous for the large animals. They break through the hardened snow, sometimes cutting their leg muscles. Rodents and other lightweight animals can run easily across the crust, but the firm surface also makes hunting easier for owls, one of the rodent's natural predators. A shrew foraging for insects on the crusty snow makes more noise than it would on soil: it may be unable to tunnel to safety fast enough to avoid the owl's sensitive ears and sharp talons.

Some birds, such as the willow ptarmigan, take advantage of snow's insulation by burying themselves at night. When snow is soft, the ptarmigan flies in at full speed; in harder snow it tunnels with its claws.   [14]

PLANTS, GARDENS, & TREES

WILDFLOWERS: MIRACLES OF SPRING

Blossoming wildflowers are one of the great joys of springtime. Every year we are surprised again by the fresh radiance of the colors and the sweetness of the fragrances. What a miracle! Why are we treated to such a glorious display each spring? Of course, the answer is that spring is an ideal time for plants: the ground is swollen with moisture, and the temperature is just right for efficient photosynthesis, so plants can rapidly transform light energy into new growing tissue.

As soon as it is warm enough to function, plants are in a rush to get going. Every plant must build itself from "nothing," starting with only the meager photosynthetic production of those few first budding leaves. In deserts and cold environments (high mountains and extreme latitudes) there's a rush to produce flowers and fruits during the short growing season. In all environments the name of the game is grab space before someone else does.

The timing of first leaf production is critical: too soon and there is danger of frost-kill; too late and your neighbors have shaded you out. Sprouts from seedlings are in the most fragile situation, since a tiny seed spends all its stored energy on those first leaves. Various underground structures, such as bulbs and tubers, hold large quantities of stored carbohydrates which enable plants to get an early start in the spring. Perennials, plants which live for several years, store carbohydrates in their roots and stems to provide for their first leaves.

An amazing aspect of plant physiology is that they are so precisely adapted to the climatic regime of their immediate environment that they "know" when it is safe to initiate growth. Each local race of each species has a specific and complex set of "programs" for germination, leaf production, and flowering which respond to factors such as day length, moisture, and temperature. Plants respond to variable environmental contingencies through a complex of chemical feedback systems which trigger or suppress biochemical activity.

Another reason plants leaf and bloom at different times is that they vary in their cold tolerance. Many early-sprouting plants avoid the cold by producing their leaves in tight radiating clusters flat against the ground where it is warmer than in the surrounding air. Some of these plants produce their flowers and fruits at ground level too, while others send up a tall flowering stalk once the weather warms up. Woolly hairs on leaves act as insulation; tiny leaves are less prone to freezing; and chemical cell constituents (comparable to antifreeze) help in cold resistance. [1]

RED OSIER BRINGS COLOR TO MARCH

In early spring, my eyes become hungry for color. I am tired of whites, browns, and grays, slush, melting snow, and dead vegetation. As I sweep the fields and roadsides in search of something for my eyes to settle on, I notice a bright red brushy growth that looks somewhat like berry canes. It's been there all along, but the slender stems suddenly seem to be getting redder. When the greens come, this red will settle back into the landscape, but right now it's all we've got. March is the month of the shrubby dogwood called red osier.

When I think of a dogwood, I think of a small tree with beautiful white or pink flowers. This flowering tree, with its showy bracts that look like four large petals, is probably the best-known of the North American dogwoods, but there are several other members of the family. Most of them are shrubs rather than trees, and one of them, the bunchberry, is so small that it seems more like a woodland wildflower than a shrub. The distinction between a tree and shrub is one of size and growth habit rather than evolutionary history, and therefore all the North American dogwoods belong to the same genus—*Cornus*. The trees are at least 10–15 feet tall, with a single stem and a well-developed crown, whereas the shrubs are shorter, with several stems growing from a clump. Red osier (*Cornus stolonifera*) is a small shrub, generally shorter than I am. In size it is closer to berry canes, meadow-sweet, and steeple bush than to another familiar shrub, the lilac.

Red osier grows in wet places—along banks of streams, in ditches beside country roads, in low-lying pockets in the medial strip of I-89, and in soggy meadows that haven't been mowed for a while. Once I noticed it, I began seeing it everywhere. The red is indeed eye-catching. It's the color of the lipstick I used to think was attractive when I was a teeny-bopper back in the 1950s—a brazen purple-red. The "osier" part of the common name comes from this dogwood's resemblance to the willows used in making baskets and wicker furniture. Both the willows and their straight, pliable twigs are called "osiers."

A dogwood is distinguished from a willow by the arrangement of its twigs and buds. Whereas a willow's are positioned alternatively—first on one side, then a ways up on the other—a dogwood's are directly across from each other. Dogwood is the "D" in the nonsense phrase "MAD CAP HORSE," which helps forestry students remember that Maples, Ashes, and Dogwoods; a family of shrubs called the CAPrifoliaceae; and HORSE chestnuts are opposite. Everything else is alternate.

Besides offering human eyes a touch of color in winter and early spring, red osiers offer wildlife food during other seasons of the year. White-tailed deer, cottontail rabbits, and snowshoe hares like the twigs; ruffed grouse and wild turkeys like the buds; and wood ducks, many songbirds, and mammals like the small cherrylike white fruits. Another important role played by red osier is erosion control. Because it can endure water and doesn't even seem to mind being submerged for a while in the spring, its shallow roots hold on to soil in wet places. The plant spreads by means of horizontal stems called stolons, so gradually a brushy thicket develops where one red osier has gotten a start, and the soil becomes increasingly secure.   [2]

SPRING FLOWERS

Many of us feel a sense of joy every time we see the first of each species of the spring flowers. We want to say, "I know you, old friend, and I haven't seen you in such a long time." Spring flowers are also a treat because they come up before mud season has completely left us and before the leaves have completely emerged on the trees. The beauty of the hepatica with its pale-blue, purple, pink, or white flower can be matched only by the Dutchman's-breeches, which almost look like toys; or the spring beauty with its delicate pink stripes; or bloodroot with its glistening white flower and deeply lobed leaf uncurling from around it; or . . . Well, there seem to be many flowers just as beautiful.

The plants use a great deal of energy to produce their lovely flowers, not for their benefit, but for the plant's benefit; for the flowers exist solely for the purpose of producing the seeds that will in turn produce new plants. The beautiful petals are designed to attract not our gaze but the gaze of an insect such as a bee.

Showy flowers attract bees in order to employ them as pollinators. Pollination is an essential step in the production of seeds, for it brings together the pollen and egg cells. Most flowers have male reproductive structures found on the stamens and female reproductive structures found on the pistil. Powdery pollen comes out of the little sac, the anther, which is found at the end of the stamen. The pistil consists of three parts: the ovary; a slender style rising from the ovary; and a sticky stigma at the top of the style.

Many flowers have methods to discourage self-pollination. For example, the stigma often is not receptive until most of the pollen has left the flower. In addition, a flower's stamens are usually positioned lower than the pistil.

Insects attracted to the flower inadvertently carry pollen to the other flowers of the same species. The pollen grains germinate in the sticky, sugary solution on the stigma. The grains burst their coats and send long pollen tubes down through the style into the ovary. When this happens, the tiny sperm cells are released in the pollen tube. Each fertilized ovule becomes a seed. That same seed the next year may develop into a plant which will have a flower and will attract insects and people by its beauty.   [3]

SPRING IN THE GARDEN

The peas are up and the lettuces and chard; the spinach is loving this damp spring weather. The onions stand at attention like a little green regiment and the broccoli plants emerge above their collars of cut-off milk cartons. We're going to foil those horrid cutworms! The carrots are planted too and have been given the hot water treatment; boiling water poured on the seeds as they lie in their rows before covering with earth. Well, we'll see.

Early this year, when the seed catalogues came and brightened up a cold January, we decided to try some new vegetables, some of those cabbaggy ones from Japan and China. We wonder how *Mizuna*, a Japanese variety, is going to like it here in Vermont.

And once again we'll try to raise eggplant. A few years ago we had so many that even we got tired of Parmigiana and moussaka. But last year an eggplant blight fell upon our garden. The plants sulked, dropped their leaves, and finally, with great effort, produced three minuscule nubbins about as big as a grape. We had done everything we were supposed to, manurey compost and water when it was needed, but it must have been that our New England weather wasn't what these Mediterranean vegetables were looking for. We're stubborn, though; we'll try them once more.

We have a pile of lovely old cow manure just waiting for our tomato plants to be set into. What a feast they'll find waiting for them in their holes! And to make sure that some Zone 4 weather doesn't discourage the growth, we'll set our *Wall O Water* shells about them.

The squash, beans, and cucumbers are still in their packets, but at the end of May they'll go in with the rest.

Though the garden looks neat and tidy, we know that the weeds are just waiting until we turn our heads, the woodchuck family is licking their lips at the thought of the peas, and just wait until the cabbage butterfly spots those broccoli plants! But somehow there always seems to be enough to go around at the end of the season.   [4]

SEED DISPERSAL: INGENIOUS WAYS TO GET AWAY

Anyone who has blown the fluffy seeds from a ripe dandelion or tossed an apple core onto the ground has unwittingly contributed to one of the most important missions in the plant world—seed dispersal. For without the dispersal of seeds to new locations, young seedlings would be competing with their parent plants, often unsuccessfully, for sunlight, soil, water, and nutrients, and the plant's success as a species could well be endangered.

Seed production and dispersal may not seem especially significant to those of us whose favorite part of a plant's life cycle is the flowering stage, but for the plant it is the ultimate goal. Flowers are just one step in the process; they are the plant's way of conceiving, fertilizing, and nurturing the tiny plant embryos as they develop into seeds.

Seeds are well adapted to house the plant's next generation because they provide both nourishment and protection for the infant plant. An inner layer surrounding the embryo stores enough food to nourish the tiny plant when it first sprouts until its roots can take nutrients from the soil and its leaves can produce their own food.

The outer seed coat protects the embryo from drying out, freezing, and being destroyed by some animals. An apple seed is apt to be eaten, but its seed coat is relatively smooth and hard, so it passes through an animal's digestive system intact. Each kind of seed, no matter how tiny, has its own distinctive seed coat.

The formation of viable seeds is a plant's primary goal; their dispersal to a favorable location is the next assignment. Plants don't move, so how can seeds travel? Among flowering plants, it is at this stage that the seed container plays a vital role, whether it be an apple, an acorn, or a coconut. Plants package their seeds in whatever way best guarantees dispersal. Biologically, the seed is a fertilized, ripened ovule and its container the ripened ovary.   [5]

THE GROWTH OF A SEED

People who plant gardens anxiously await the growth of their seeds. Miraculously enough, seeds placed into the earth develop into entire plants. We all know this, and we all know some of the basic components necessary for plant growth—sunlight, carbon dioxide, and water. How plants actually grow is less commonly understood.

Trying to decide where plant growth starts is like trying to decide which came first, the chicken or the egg. Let us begin with the fertilized plant egg cell. This cell undergoes successive divisions to produce a blob of undifferentiated cells. These cells then differentiate into an embryo containing a tiny seed tip, one or two seedling leaves, and the beginning of a root. The seedling leaves, or cotyledons (in the case of dicots), take up the bulk of many seeds, since they have developed into food-storage organs for the embryo. All this, enclosed by a seed coat, is known as the seed.

After planting our seeds, we make sure to water the soil. This water is necessary for the first stage of germination, which is swelling, a taking in of water by the whole seed. Once each cell has absorbed the adequate amount of water, the embryo can actually start growing. The cotyledons deliver a plentiful supply of sugar to the stem and root cells. Respiration increases, and energy is released. Enzymes digest starch into sugar. Water continues to be absorbed. Because of the design of the cell wall, the root and young stem elongate instead of growing indiscriminately. Each cell becomes 10 to 100 times longer.

In order for there to be further growth, new cells are required. It is at the stem tip and root tip where this cell division occurs. As these special cells on the growing tips divide and multiply, they leave behind them a zone of elongation. As these cells mature, they form a variety of different tissues needed by the plant. The root is the first part to emerge from the seed. It probes the earth for water. The sprout then emerges and pushes its way up through the soil. The seedling is on its way to becoming an adult plant. [6]

SOIL

Soil is a complex mixture of inorganic matter, decayed or decaying organic matter, water, dissolved minerals, and air. It provides plants with a place to anchor themselves, a reservoir of nutrients, and a transportation network for water and air. Soil is said to have texture, the technical term for particle size, and structure, the technical term for the grouping or arrangement of the particles. Together, the texture and structure of soil determine its heat capacity, aeration, its ease of tillage, its fertility, and its ability to hold and conduct water. While soil texture is an unchangeable physical property, soil structure can be improved by the addition of organic matter or humus (completely decayed organic matter).

Soil used for farming can rapidly become depleted when harvesting the crop prevents the natural recycling of nutrients. The nutrients lost by removal of the crop must then be replaced with fertilizers. To be complete, a fertilizer must provide the plant with thirteen essential minerals. Nitrogen (actually not a mineral), phosphorus, potassium, calcium, sulfur, and magnesium are called the major elements because they are needed in large quantities; iron, boron, copper, manganese, molybdenum, zinc, and cobalt are called the trace elements because they are needed only in small quantities.

Whether to provide the necessary nutrients in the form of a chemical fertilizer or an organic fertilizer is the subject of heated debate at times. While there is no chemical difference between the nutrients provided by chemical fertilizers and organic fertilizers, there are other dissimilarities. Since organic fertilizers contain a more complex mixture of molecules than chemical fertilizers, they supply smaller doses of nutrients over a longer period of time; this eliminates the danger of "burning" plants by providing too concentrated a dose of nutrients but risks not providing enough nutrients. Since chemical fertilizers are more concentrated, they are easier to handle than organic fertilizers. Chemical fertilizers are often less expensive; there is potential for rapid price increases, however, because their manufacture consumes large quantities of energy. The extra bulkiness of organic fertilizers (organic matter and humus) improves the soil structure, although organic matter may harbor weed seeds, pests, and parasites capable of damaging the crop. To improve soil structure when using chemical fertilizers, peat or vermiculite must be added along with the fertilizer in some cases.

The dilemma about types of fertilizers is further complicated by the fact that organic fertilizers rarely, if ever, provide enough nutrients to produce the maximum possible yield and do not necessarily contain the full complement of nutrients. Yet organic fertilizers, by improving soil structure, help prevent erosion, one of the prime factors in nutrient loss and possibly the most important factor in limiting long-term soil fertility.   [7]

PLANTING TREES

Over the last few years we have been searching for clues to the declining health of community trees, and we are coming to believe that planting methods are a major culprit. The main reason is that home construction has changed greatly since the good old days. Bigger earth-moving equipment and less hand labor are used in creating today's housing developments. Because of the heavier construction equipment, the soil in the average yard is less fertile and more compacted.

Digging a hole in dense, compacted soil and filling the hole with peat moss and other soil amendments is like creating a pot for the tree. The roots grow outward in the soil amendments, and the tree does fine until the roots reach the original soil and the outward growth stops. Instead of spreading into the yard, the roots encircle the planting pit. The "pot" soon fills with roots, and the health of the tree declines.

The crown continues to grow, but the roots do not. Once the tree becomes root-bound, its ability to maintain itself during a drought or to survive a flood is limited—leading to decline that is often terminal.

So what do we propose? Plant so that roots have a chance to grow into the surrounding soil and produce healthy, vigorous branches, foliage, and roots. Instead of a planting hole, what's needed is a large planting area that is wide but not deep, where the soil is loose and suited for root growth. The larger the area, the better.

After selecting a suitable location, mark out a planting area that is five times the diameter of the planting ball. Use a rototiller or shovel to loosen and mix the soil in this entire area to a depth of about twelve inches. Organic matter can be added to the loosened soil so long as the new material is used uniformly throughout the area.

In the center of the prepared area, dig a shallow hole to set the tree, root ball and all. The hole should allow the root ball to sit on solid ground rather than loose soil. Once the ball is set in the hole, its upper surface should be level with the existing soil.

After the tree is properly situated, cut and remove the rope or wires holding the burlap in place and securing any part of the tree. Position the tree so that it is perpendicular to the ground, so the main stem will grow straight up.

Backfill around the root area and gently step the soil to prevent major air pockets, but it is a mistake to pack the soil too hard. Water can be used instead of your foot to help the soil settle and prevent overpacking. Rake the soil even over the entire area and lay mulch on the area using two to four inches of bark, wood chips, old sawdust, pine needles, leaf mold, or the like. Some mulches decompose quickly and will have to be replenished once or twice a year. Maintaining the mulch layer carefully will improve tree growth substantially.

[8]

LEAFY WONDERS

I think that I shall never see,
A poem lovely as a tree.

These familiar lines from Joyce Kilmer's tribute to trees capture the way many people feel about these remarkable plants. Trees are the most massive of all living things, with the giant sequoias of California holding the record. The General Sherman tree is the largest, with a height of 275 feet and a girth of 32 feet. The world's tallest trees, the redwoods, are also native to California. Slimmer than the giant sequoias, these graceful trees grow to heights of 275 feet and more. The tallest on record soars more than 360 feet above the earth.

Trees are also among the oldest of all living things. The ancient bristlecone pines in the Sierra Nevada mountains of California are more than 4,600 years old, about the same age as the pyramids of Egypt.

While the size and age of trees are remarkable, there are many far more important characteristics that make these plants essential to us. Perhaps the most important is the production of oxygen. As a by-product of photosynthesis, the leaves convert carbon dioxide to oxygen. While all green plants duplicate this process, trees produce the bulk of the oxygen we breathe.

Trees also provide us with many other products like lumber, pulp for paper, turpentine, cork, rubber, nuts, medicines, and fruit. On a hot day, the cooling shade of trees gives relief from the sun. Large forests even help moderate the climate of the earth. Trees are essential in controlling the erosion of topsoil. The root systems not only help hold the soil in place but also absorb runoff that would otherwise wash the soil away. Even after death, trees are still an essential part of nature. Their rotting hulks provide homes for countless creatures, and eventually the decay process returns nutrients to the soil to be recycled.

When we look at a tree, it is the height that charms us, and we think nothing of the roots; but the tree could not be without them. The roots provide support and absorb water from the soil. Water is taken in by the fine hairs that grow near the root tips. It then passes into the central woody portion of the root, which carries it to the stem and leaves. The roots of some trees reach astoundingly deep. The taproot of an ancient hickory or oak may bore down as far as one hundred feet. [9]

PAPYRUS

We saw some of it while on the Nile in Egypt. Much of it
has disappeared. For the first four-fifths of its history, the
material on which Western civilization did most of its writing
was not paper but papyrus. Rag paper was invented by the
Chinese. The Arabs learned the manufacturing process from
the Chinese and introduced the product to the Mediterranean world, where it came into common use after A.D. 1000.
Paper therefore tells us nothing about writing in antiquity.
For this, one has to turn to the rush-like plant that gives
papyrus its name—*Cyperus papyrus* in Linnaeus's taxon-
omy—a versatile plant still growing in many parts of the
world.

It is the papyrus of ancient Egypt that concerns us here:
all through ancient times, it grew luxuriantly in the marshes
along the Nile, its stalks sometimes as high as fifteen feet;
these are the "bulrushes" in which the infant Moses was
found. The Egyptians exploited every part of this valuable
plant; they burned its woody rhizome as fuel and also used
the rhizome in the manufacture of furniture and utensils.
With its feathery crown they made floral offerings and ornaments. Even more vitally, they used its stalk as food—in the
words of one ancient writer, as a "refuge from want for the
poor." They boiled or baked the stalk as we do vegetables,
and they also ate it raw, chewing it for the sweet juice and
spitting out the pulp, much as is done with young bamboo
shoots. Stalks of papyrus were lashed together to make boats,
the seaworthiness of which was demonstrated a few years
ago by Thor Heyerdahl's transatlantic "Ra" expeditions. And
with strips of the tough, fibrous cortex of the papyrus, the
Egyptians wove plaited products—mats, baskets, sails, rope,
candlewicks, and even sandals, examples of which, decorated with brightly colored beads, have been found in Egyptian tombs.

But most important in the history of civilization was the
Egyptians' discovery, sometime prior to 3000 B.C., that a
writing material could be readily manufactured from papy-

rus. Thin strips were cut from the pith of the stalk and laid side by side in two transverse layers; these layers, pressed together, adhered to form a surface eminently suitable for writing. Soot mixed with water served as early ink, and the earliest pens were reeds sharpened to a point.   [10]

GRASSES: SLENDER STALKS WITH SEEDS THAT NOURISH THE WORLD

Measured in terms of geographic distribution and numbers of individual plants, the grass family is the most successful flowering plant family in the world. The almost five thousand species, about one-third of which are found in North America, are distributed from tropic to tundra and from marsh to desert. Only the orchid and composite (for example, sunflowers, daisies) families have more species, but neither is as numerous in individual plants or grows in such diverse and widespread areas.

One key to the success of the grasses is the simplicity and adaptability of their basic design. The root system is dense and spreading, enabling the plants to utilize all available moisture, an important adaptation for plants that frequently grow in dry areas. This root mat also enables grasses to hold on to loose, sandy, or muddy soils such as marshes and sand dunes, where they are often found.

Rhizomes, or underground stems, are an intricate part of the structure of many grasses. Like roots, rhizomes are important, in that they help grasses hold on to soil. A second important function of rhizomes is propagation. These underground stems send up numerous shoots. If you have ever tried to eliminate witchgrass from a garden, you are familiar with the white runners that send up what seem like endless plants. Rhizomes are an important design for the survival of grasses in areas that are burned frequently, such as prairies. While surface vegetation is burned, the underground stems are unharmed and quickly send up replacements. This adaptation is one of the factors that enable grasses to outcompete trees in prairie areas, where lightning fires are common.

The stems of grasses are joined with elongated round hollow sections interspersed with compact solid sections called nodes. Unlike most other plants, grass stems grow from the nodes, not from the tops or ends of branches—a necessity for a plant family that is browsed by a wide variety of animals. The nodes help provide additional rigidity for the plant as well as acting as points from which the leaves originate. The leaves encircle the stem, one above each node, in a sheath, and then protrude in different directions to provide for maximum exposure to the sun. Clasping the stem affords the long, narrow leaves much-needed support. Some stems grow vertically; others may trail along the ground and root anew at each node.   {11}

DANDELIONS: SURVIVORS IN A CHALLENGING WORLD

Dandelions may well be the first flowers that many of us learned to call by name. Their abundance, their relatively early blooming, and their use/nuisance reputations bring them to the attention of children and adults alike. The name dandelion is also easy to remember, especially if you know its derivation, "dents de lion," or teeth of a lion.

Dandelions are members of the composite family, which gives them a number of famous relations, such as daisies, hawkweed, sunflowers, chicory, goldenrod, burdock, and aster, to name a few. Among flowering plants, the composites are second only to the orchid family in number of species, with something over thirteen thousand species. The dictionary defines composite as being made up of distinct components. This relates closely to the botanical meaning of the word *composite*, which describes flowerheads composed of many tiny individual flowers; when you pick one dandelion flower, you actually pick approximately three hundred dandelion flowers or florets.

People familiar with dandelions probably recognize the long, deeply indented or toothed leaves, the round, compact, fuzzy-textured yellow head, and the ball-shaped, white, symmetrical seed head. How do these three familiar parts connect with one another to create a successful plant? During the first year the leaves, which grow in a circular pattern close to the ground, are the only visible part of the plant. These first-year leaves make food that is stored in the long carrot-shaped root, thus enabling the plant to get a head start for the second and succeeding years in making flowers.

One dandelion plant can produce many flowerheads at a time and in rapid succession. The unopened flowerhead is protected by green bracts folded up around it; these same bracts curl down when the flowers emerge but are ready to shutter in the flowers on rainy, cloudy days and at night. When a dandelion flowerhead blossoms, the individual florets, each a complete miniature flower, mature in circular rows starting at the outside rim. First, pollen is produced from the anthers, which are fused together in a tube-like formation around the stigma (this is very hard to see, even with a hand lens). Then the female part, the stigma, pushes up through the pollen, carrying pollen grains with it.

Dandelions are rarely pollinated by insects, as one might think; the many insects that visit dandelions gather pollen, but they only cross-pollinate one in ten thousand dandelions. Rather than forming as a result of pollination, dandelion seeds usually form from a part of the parent plant. Dandelions are one of the few plants in which seeds can form without pollination occurring; in effect, many dandelions are thus clones of their parent plant. One can see variations in dandelions, but those growing close together look very much alike. [12]

GOTCHA!

When we think of predators, we usually think of animals that go out and hunt for their food—animals such as wolves, hawks, and snakes. But certain animals do the opposite; they stay put and trick their prey into coming to them. Plants do a similar thing when they lure birds and insects to help spread their pollen or seeds. Since plants can't move around, this type of trickery isn't surprising. What *is* surprising is that some plants actually trap and eat the animals they attract.

It's easy to understand why animals eat other animals. To stay alive, they need the carbohydrates, fats, proteins, vitamins, and minerals contained in their prey. What about plants? They generally make their own food out of sunlight, carbon dioxide, and nutrients found in the soil. But not all soils provide the full range of nutrients a plant needs. Plants that live in the acid soil of some bogs and swamps have to look elsewhere for nitrogen, an important nutrient. Insects are rich in nitrogen, so these plants have developed ways of capturing them.

Plants that don't have chlorophyll, which is used by other plants to make food, face a similar problem. They have to get their nutrients from other living things, often by attaching themselves to other plants. But there's one type of fungus that traps microscopic worms for food.

Of course, plants don't actually eat their prey the same way animals do. Instead of chewing up the insects they trap, they "swallow" them whole and then digest them, much as a snake's stomach and intestines digest a mouse. Some plants produce their own digestive juices; others let bacteria do the work and then share in the meal.

Plants and animals capture their prey in different ways. Some plants lure an insect with sweet-smelling nectar, then hold it fast with a sticky substance or trap it in a liquid-filled chamber. Others, such as the Venus's-flytrap, spring shut around an insect when it touches a part of the plant.

Animals such as spiders and caddis fly young (called larvae) spin webs to catch their prey, while the ant lion digs a pit and waits at the bottom for something to fall in. Flashlight fish attract their prey with beams of light, and the sargassum fish uses its dorsal fin as a fishing lure.

Only a few plants that capture insects end up eating them. Most simply hold the insect long enough for any pollen it has brought with it to fall off and for the capturing plant's pollen to stick to its body. This can take anywhere from a few minutes to a couple of days. Some plants actually feed their catch so it will be strong enough to escape and pollinate other plants.

The idea of a trap is simple enough, and humans were quick to adopt it. But traps made by people are crude compared to those found in nature.   [13]

SILAGE

Silage is pickled corn plants, made in, naturally enough, a silo. Silage-making is a little like home brewing of beer—no two batches come out exactly alike, and yeasts produce a number of acids which cause the final product to vary some. Changing the ingredients can also vary the product, and pickled hay (most grasses can be ensiled) often has additives salted into it to control the outcome. However, nothing is needed to produce *corn* silage; all the necessaries are there in the plant itself.

The term *fermentation* has been used for four hundred years to describe decomposition processes in food where a gas is involved. Brewers have known how to do it since the

1600s, but it was 1837 before fermentation was understood and its causes correctly discovered. Here is how the chemistry proceeds deep in the darkness of the silo.

Whole corn plants are chopped into little pieces—stalks and leaves and ears—everything but the roots. These bits are sucked up by a blower and hurled into the top of the silo (if it is the familiar vertical type) or pushed by a loader into the bunker silo (if it is a horizontal type). As the corn bits are piled on, their own weight packs them down, squeezing out air from within the plant cells and from between the corn bits. The corn has been freshly cut, and respiration continues within the plant, giving off carbon dioxide and raising nitrogen levels and the temperature of the silo.

Bacteria, yeast, and mold (found everywhere—on the corn, in the air—everywhere) begin to grow, feeding upon the carbohydrates in the corn, converting them to acids. Plant enzymes also continue active for a while, tearing down sugars into alcohol and acid and water, reducing proteins to amino acids and ammonia. With all this heavy breathing in the cells, oxygen is soon depleted in the silo; oxidation ends, and the anaerobes awaken. The yeast grows for a time but soon goes inactive, leaving the whole place to the bacteria. In time, *they* foul their own nest too, creating conditions so acidic that they cannot live. When this toxic level of pH is reached, when the anaerobes die off, the stew is done. The corn is pickled and will stay that way, even after it is exposed to air again. [14]

FASCINATING FUNGI

There are those things in nature that elicit wonder. There are those that elicit dread. And there are those things that elicit continual scientific consternation. Perhaps no group of organisms is better characterized by a combination of all of these qualities than the fungi.

Mushrooms have fascinated and intrigued mankind for centuries, and for no less time have they been subject to many a myth or folktale. It has been said they mysteriously arise from the foaming froth of furious stallions in the sky or from the gentle footsteps of dancing hobgoblins. They have been acclaimed to hold mystical powers and have served as the spiritual focus of shamanistic cults all over the world. The hallucinogenic powers of certain species have inspired visions that have both created and revolutionized religions. Used incorrectly, however, certain species can kill a human being in a matter of minutes.

Indeed, mushrooms have captivated us throughout history. What else could one expect from an organism, some species of which can turn blue at the touch of a finger or produce hundreds of fruiting bodies in a couple of hours or rot the timbers of a person's house or strangle and digest microscopic nematode worms? What we do know, however, is that:

1. A fungus is primarily composed of an underground network of rootlike hairs called a mycelium, which serve to nourish the organism throughout the year by the absorption of various nutrients from the substrate.

2. While the mycelium may lie dormant during months or even years of dryness, cold, or heat, at a certain point when the climatic conditions are just right (i.e., proper water moisture, correct night length, increase in CO_2 levels), it grows vigorously enough to produce a spore-bearing fruiting body or mushroom, much the way a healthy plant will produce a flower.

3. A fungus tends to be parasitic, saprophytic, or mycorrhizal—that is, it derives its food source from a special relationship with the roots of certain vascular plants.   [15]

THE MOREL MUSHROOM

Morchella esculenta—even the name sounds like a delicacy. For centuries people have combed the woods and fields each spring in search of the morel mushroom, one of nature's most tantalizing treats. The taste defies description, but if price is any indication of quality, the morel is the Cadillac of mushrooms. Fresh morels cost from $7 to $10 per pound, and, in the off-season, dried morels sell for $50 to $200 per pound.

Actually, there are many different species of true morels. They are all edible, and each species has its own unique flavor. Our South Dakota listeners tell me they have three or four species, but *Morchella esculenta*, the "yellow morel," is the most common and one of the most tasty.

All true morels are distinctive in appearance and typically grow four to eight inches tall. On the northerns plains they are almost always restricted to forest and woodland habitats. The forests along rivers and streams seem to be the most prodigious producers of morels. The morel season is short; it usually lasts for about three weeks, most often in May. The precise timing depends on the weather, but some say that the morels are out when the oak leaves are the size of squirrel ears. When the lilacs are in bloom, it's time to look for morels.

Botanists have tried for centuries to propagate morel mushrooms artificially. What we see growing above ground is actually only the fruiting body of a plant that lies mostly below the ground surface. The subterranean portion of the morel consists of a tangled mass of fine fibers called "mycelium." Botanists have been able to grow the mycelium in the laboratory but have not been able to stimulate "sporulation."

Well, the morel's secret has finally been discovered. In the early 1980s, Ron Ower was a graduate student at San Francisco State, and it is he who discovered a technique to make the morel mycelium sporulate. Commercial development of this technique is underway, and the first artificially produced morels reached the markets of Switzerland in 1990.

A word of caution: there is a group of plants called "false morels," which superficially resemble the true morels. And some of the false morels are poisonous. When it comes to distinguishing edible mushrooms from poisonous ones, there is only one simple rule: If you can't positively identify it, don't eat it!   [16]

LICHENS

Lichens can be called the pioneers of the plant kingdom. They will grow practically anywhere: on dry, bare rocks, in soil, on tree trunks, and on the arctic tundra. Lichens are usually the first plants to grow in a new land environment.

One of the most fascinating facts about lichens is that they are really two plants living together: an alga and a fungus. The algae would be unable to survive in a harsh, dry environment, and the fungi, having no chlorophyll, are not able to manufacture their own food. Together, in a lichen, the fungus provides support, moisture, and minerals for the alga, and the alga provides food for the fungus. Such a relationship, by means of which both partners benefit, is known as symbiosis.

The shape of a lichen depends on the structure of the fungus component. The ten thousand species of lichens are classified into three main groups, based on their general structure. Crustose lichens are flat and crusty and grow on rocks or tree trunks, sometimes becoming embedded in them. Foliose lichens are not as firmly attached; the free edges are often leafy or lobed in appearance. Fruticose lichens usually stand upright and have a branched structure. The most familiar fruticose lichen is often mistakenly called reindeer moss. It has a branched structure, like antlers, and is an important food for reindeer, caribou, and other arctic animals.

Lichens absorb moisture directly from the air. They are therefore very sensitive to small amounts of impurities and are good indicators of air quality. Polluted city air discourages lichen growth.

Lichens are very slow-growing; you can assume that a rock with a large lichen growing on it has been there for a long time. As the lichen grows, it produces chemicals that slowly break down the rock. The bits of eroded rock and decayed plant material gradually form soil. After many years, plant succession begins with the establishment of mosses, followed in later years by ferns and small seed plants. After several hundred years, a forest may grow where there was once only lichen-covered rock. [17]

IN DEFENSE OF DEADWOOD, DECADENCE, AND DISORDER

Consider prevailing attitudes toward dead and dying trees. They have little or no commercial value. Occasionally they shade potentially valuable growth. Perhaps they harbor disease. Industrial foresters tend to look upon them as untidy, in need of surgical excision.

Environmentalists, on the other hand, value dead and dying trees because they shelter and feed countless wildlife species. Such "den trees," however, are in critically short supply throughout most of the nation. "There is a definite shortage of den trees, and it's becoming more acute all the time," says Henry Laramie, game supervisor for the New Hampshire Fish and Game Department. Laramie reports that little is being done to save the few den trees that still survive.

The history of a large dead or dying tree reads like a wildlife field-guide. Insects come first, even before the sap has hardened—borers, beetles, the "grub" larvae of a thousand flying forms. Then the woodpeckers, the smaller of the tribe ventilating bark—the larger, led by the jackhammer-billed pileated woodpecker, excavating five-inch-deep termite mines in the still-fresh wood. Water finds its way into the holes and speeds the rotting process. Fungi side-step their way up the trunk. Gray squirrels eat the fungi and nest in the woodpecker holes. Flying squirrels eat the insects and nest in the woodpecker holes. Hawks and great-horned owls eat the squirrels and nest in the dead branches. Each year there are new woodpecker holes; and the old ones get larger, housing new forest creatures. One year, screech owls, the next a pair of wood ducks from a beaver pond a mile west, the next a pair of hooded mergansers, the next a pair of pileated woodpeckers—who started it all. Opossums, raccoons, red squirrels, porcupines, gray foxes, fishers, bobcats, white-footed mice, big brown bats, barred owls, saw-whet owls, tree swallows, nuthatches, chickadees, wrens, titmice, bluebirds, goldeneyes, kestrels . . . Gradually the creatures who use a dead or dying tree return it to the forest floor, where it nourishes a seedling that will, if permitted, grow and die, shelter and feed new generations of wildlife, then return to the earth to nourish another seedling.

The fact that many wildlife species are now scarce—or at least scarcer than they used to be—is due in large measure to the shortage of den trees. [18]

GALLS: SMALL HOMES FOR TINY CREATURES

What are those odd swellings on some plants, the lumps and bumps on certain leaves? Look closely—perhaps they are galls. A gall is an abnormal growth of plant tissue produced by a stimulus external to the plant itself. Stated more simply, some substance injected into a plant causes that part of the plant to swell or grow in a particular pattern. It may be on the leaf, stem, flowerheads, stalk, or root, but it can develop only while that part of the plant is growing, which explains why galls usually form in the spring.

Galls come in many shapes: round, conical, kidney-, disk-, or spindle-shaped. Their textures vary greatly too, from almost fluffy to papery or woody, and from smooth to sticky, hairy, bumpy, scaly, or ridged.

The purpose of a gall is to provide a home for a tiny creature. And what an incredible home it is, with its solid outer walls and constant food supply. Insects and sometimes mites are the chief tenants of galls during the immature stages of their lives. Besides producing food and shelter, the gall keeps its occupants comparatively safe from parasites and predators, and it protects them from drying out.

Exactly what initiates the gall growth on the plant is not clear. It appears to be a chemical substance either injected by the mother insect at the time she lays her eggs or secreted in the saliva of the larvae as they bore into the plant for shelter or begin eating it for food. Sometimes the adult female seems to start the process and her offspring, carrying precisely the same chemical, continue to stimulate the growth of the gall.

How galls form and grow on a plant is intriguing. The initial stimulant causes starch to convert to sugar, resulting in an excess of food material, which stimulates the plant cells to enlarge and/or multiply and the gall to grow. Usually the presence of a gall does not harm the host plant. Although galls grow on a great variety of plants, oaks are the most popular, claiming 800 of the 2,000 kinds of American galls, followed by willows, poplars, plants that belong to the rose family, such as blackberries and raspberries, and finally composite plants such as goldenrods and asters.   [19]

MANGROVES

Although the Caribbean shore is more famous for its coral sand beaches and wave-splashed rocks, nearly 25 percent of its coastline belongs to the mangroves. The quiet places, the still waters of the estuary or lagoon that attract more herons and pelicans than people—those are the places where the graceful prop roots of the mangrove trees arch seaward into the shallows. As the mangroves grow outward from the shore, they gradually stabilize and build the coastline. Leaves drop, silt gathers, the shells of sea animals and coral rubble compound. Almost imperceptibly the space beneath the roots fills, and the shore bows outward.

The activity is more visible in the canopy of branches overhead and in the interlacing roots of the trees. Flocks of storks, herons, egrets, cormorants, kingfishers, and ibises may roost on branches where lizards sun, snails cling, and crabs scuttle down toward the water. At the waterline the mangrove roots are thick with flat, dull clusters of mangrove oysters. When Sir Walter Raleigh returned to England after a 1595 visit to Barbados, he was nearly laughed out of court for his description of these oysters that apparently grow on trees.

The mangrove ecosystem is just as unique beneath the water, where the looping roots interlace along the sea bottom. The newest roots hang short and wave with the currents. Often they are covered with coral, sponges, algae, and mollusks.

The still, protected waters around the roots act as a nursery for large numbers of commercially valuable fish, lobsters, and mollusks. Grunt, snapper, grouper, tarpon, and sea trout either breed or do much of their growing in mangrove areas. Conches mature in grass beds nearby. The abundance of food and shelter makes the mangrove community one of the most biologically productive of tropical ecosystems. The trees also provide natural buffer zones that absorb the shock of hurricane waves and filter pollutants from freshwater runoff.

Most plants extract oxygen from air spaces between soil particles. But in the sodden mud of the mangrove swamp, the soil is compacted and lacks air space. Mangroves solve this environmental problem by extracting oxygen directly from the air. [20]

BALM OF GILEAD

Before modern drugstores and pharmacies made medicines easily accessible to everyone, people made their own remedies from the things they had at hand and found in the woods and fields of the farm. Some of the knowledge of cures for ailments came with the early settlers from Europe and England. Many of the plants in the New World, however, are different from plants in the Old World left behind, so early farm families discovered remedies through improvisation and experimentation.

Europeans knew of the revered Balm of Gilead, even if they had never used it. The genuine Balm of Gilead tree grows in the vicinity of the Red Sea. During medieval times the Turks so valued this rare tree that they prohibited anyone from exporting it. Rare Balm of Gilead trees grew in the guarded gardens of the wealthy near Cairo, where the royal ladies valued it as a cosmetic. The queen of Sheba is said to have given Solomon Balm of Gilead trees as a present. From these trees grew the famous cultivated plantings on Mount Gilead in ancient Palestine, the reason the tree now bears this name. In fact, the word *balm* may be derived from a Hebrew word that means "chief of oils" or "a sweet smell." The resin produced by this tree continues to be valued for its healing properties.

The closest tree to this ancient, magical Balm of Gilead in the New World is the Canadian poplar, which grows throughout northern New England and Canada. Colonists soon discovered that the bark, and especially the buds, of this not-uncommon tree produced similar medicinal effects, and early in colonial history they named it Balm of Gilead after the mystical tree of the Turks.

In her family remembrances, *Out of the Salt Box* (1974), Ruth Rasey tells us how, one April evening, her great-uncle Zachariah accidentally crushed some of the fallen Balm of Gilead buds "between the top fence rail and his lumber-blistered hand" and discovered that the sticky juice exuded by the buds soothed his sore blisters. During the next few days he and his father and brothers experimented with boiling down crushed buds in the "big brass kettle slung over a fire in the stone arch at the edge of the sugar bush." They evolved a healing ointment by adding fresh hot mutton tallow to the strained decoction of Balm of Gilead buds and settled pasture spring water.

Many other settlers on rural mountain farms made various remedies from Balm of Gilead. Mixed with honey, often dissolved comb honey, and dark Jamaican rum, they produced a ubiquitous cough and cold syrup found on nearly every kitchen shelf for over a century. The buds of the Balm of Gilead are covered with a fragrant resin that dissolves in boiling water. Their odor is like incense but with a bitter, rather unpleasant taste, so Balm of Gilead syrup needs sweetening.

Besides ointment and syrup, home cures came in jelly form, produced much like any other jelly, with cup for cup of sugar added to the boiled-down juice, often dark-brown sugar containing molasses. This the farm wife put up in glass jars, and when someone in the family developed a cough, she mixed a spoonful or two with lemon juice in a cup of hot water. "It will cure any cold," one woman said. "I keep balm jelly in the house all the time." [21]

FOR WHAT AILS YOU

For as long as people have become ill, they have searched for magical cures for what ails them. When chanting and prayers and sacrificial offerings failed to contain their sneezes or soothe their aches, they turned to mystical potions and powders and healing herbs. While much of this medicinal trial-and-error was at best ineffective and at worst poisonous, there were plants and balms and teas that held healing properties. Around these strange herbs grew myth and lore which exist to this day, their powers denied by some and praised by others. Here is an introduction to the historical herb lore for some familiar plants.

The ivy, *Hedera helix*, which climbs our walls once crowned the god Bacchus, as wreathing one's head with ivy was said to prevent drunkenness. A hangover from wine could be cleared if a tea was brewed from ivy leaves.

One of the most common and widespread plants is knotgrass (*Polyganum aviculare*), also called pigweed, cowgrass, armstrong, or red robin. It has as many names as it has purported cures. This is the straggly, plucky weed growing in vacant lots that Shakespeare spoke of in *A Midsummer Night's Dream* as "the hindering knotgrass," referring to the common belief that it could retard growth in children and young animals. More recently, herbalists have used it as an astringent, a diuretic, and a way to expel gallstones when taken "one dram of the herb powdered in wine twice a day." The fresh juice is claimed to stop nosebleeds, strengthen weak joints, and comfort nerves and tendons.

The lavender, *Lavendula vera*, in your garden produces a volatile oil we notice as a delicate perfume. Moths and mice notice lavender and also mint as a deterrent, so people scatter the leaves and flowers in drawers to discourage vermin. Lavender oil is also used to restore the faint or giddy, stop palpitations and spasms, and stimulate one's appetite. A few drops are claimed to relieve fatigue, cure a toothache, sprains, rheumatism, hysteria, and palsy.

Parsley has been honored through history as far more than a garnish. The Greeks wreathed tombs with this sacred herb, and the Chinese used it to cure bladder, kidney, and prostate trouble. Most commonly, it is chewed to sweeten one's breath after eating garlic.

There are many good books on herbs and herb lore available in libraries. The next time you feel a cold coming on, why not reach for your mint and elder flower tea?

[22]

THE TREE OF LIFE

The Kiwai of New Guinea, like many preliterate groups, have a great respect for trees. In the late nineteenth century, when they were first given iron axes by the British, they were reluctant to use them; there were certain trees in particular that they refused to cut down. They believed, it was learned, that trees were inhabited by spirit beings who were somehow connected with the soul of the tree. Before any tree could be felled, the axman would have to request that the spirit move to another tree, and if his arms felt heavy while cutting or if the work seemed particularly hard, it was a clear sign that the spirit had not yet deserted the tree, and it was perhaps best left standing.

A single, good-sized maple may contain several thousand green leaves, which can expose some two thousand square yards—about half an acre—to the sky and the sun. These leaves take carbon dioxide from the air and hydrogen from the waters of the soil, combine the two with sunlight, and produce carbohydrate, a substance capable of sustaining other forms of life. A growing tree uses carbon dioxide at a rate of about forty-eight pounds per year, which comes to about ten tons for every acre of forest. Every ton of new wood extracts about one and a half tons of carbon dioxide and produces in return a little more than a ton of oxygen. If you consider that there are still about ten billion acres of forest on earth, that means that some one hundred billion tons of carbon dioxide is being used up each year by the forests of the world.

As you may know, carbon dioxide has become a problem in our time. Because of the destruction of the world's forest, because of the human use of fossil fuels and the increase in other gases associated with industrial activity, the dynamics of the earth's atmosphere and its plant life are changing, and there is now too much carbon dioxide in the earth's atmosphere. The result is a phenomenon popularly known as the greenhouse effect.

But as is often the case in world events, it is perhaps important to relearn what was known by our forebears. Trees are life, more than we may know, and as the Kiwai made clear, we should perhaps be more careful when we go into the forest.   [23]

TALKING TREES

The mystery of talking trees is being unraveled by modern science, beginning with the extraordinary research described in Peter Tompkins and Chris Bird's book, *The Secret Life of Plants* (1972). By attaching various electromagnetic devices to trees or other plants, scientists are discovering that trees not only talk to each other but listen and react to human beings. This may sound strange to modern ears, but actually people have been communicating with trees for centuries.

All over the world people have revered trees, and legends tell us of special trees that talked to particular persons in need of assistance. Trees are believed to have pronounced spells when needed or to have divulged information necessary to the listener. Trees, after all, often live very long lives by a human scale and so observe a scope of history and attain a perspective hardly possible to the comparatively short-lived person. In Sherwood Forest, for years hunters and poachers consulted a venerable elm of great size about the whereabouts of their prey. Reports on the tree's answers suggest that this mighty tree took the cagey middle road between helping the hunter and protecting the hunted. The Sherwood elm often gave enigmatic answers:

> *When the wind shakes my branches from the west,*
> *The deer will put your arrows to the test.*
> *When east winds scatter leaves about your feet,*
> *Then look to hidden rabbits for your meat.*

The Iroquois Indians, in their curing ceremony, use masks of mythical faces made of basswood that are cut from a living tree. When preparing to take the living face from the tree, the carver first offers a prayer to the basswood tree and then gives the tree some sacred tobacco so the tree will know why it is being wounded and will allow its living energy to stay in the face and aid in the healing ritual.

The living energy possessed by a tree we easily take for granted, no longer being sensitive, as were our ancestors, to the spirit of the tree. But we can learn again to hear what trees have to say by putting our ears to their trunks and listening. We can renew ourselves by sitting with our back against the tree and letting its living energy refresh us, something woodsmen and hikers have always practiced. We can then participate again in the mystery of the talking trees.

[24]

CATNIP

If you transplant it, the cats will eat it,
If you sow it, the cats don't know it.

This old saying about catnip reminds us that bruised catnip plants attract all manner of cats, who then perform the most ecstatic and hilarious antics in the process of getting their catnip high. Transplanting always bruises the plant, but catnip sown from seed we can sneak into our gardens without our cats knowing it. This way we get to enjoy the catnip too.

The leaves of catnip make a delicately delicious tea, especially if water that has come off the boil is poured over them and they steep in a covered pot. Catnip tea garnished with honey and lemon slices served after a large meal aids the digestion and helps produce the after-dinner nap, when we might be said to curl up like a contented cat. Better even than tea are candied catnip leaves. Once the ingredients are set up, they are really no trouble to make. First, we dip fresh catnip leaves in a glaze of equal parts egg whites and lemon juice. Then we sprinkle raw sugar on the leaves and allow them to dry for a day or so before eating them. A delicacy fit for both goddesses and gods.

Catnip contains a volatile oil which is easily dissipated and lost in the air if the leaves are boiled in making tea or if the plant is dried in the sun rather than in the shade when preparing the catnip as a dried herb. This oil is what attracts the cats, turning the most well-mannered, sedate cat into a ridiculous feline tumbler. Trappers in earlier times in the backwoods pressed the oil from the plant and rubbed it on their traps when setting traplines for bobcats or lynx.

The catnip, with its square stems (like all the mints) and its pale lavender to pale pink flowers dotted with small purple spots, is such a familiar plant we forget it is an immigrant to North America. Its native home stretches across Europe and Asia, and so colonists from many countries brought catnip with them, both for themselves and for their cats.

An old-fashioned way of using catnip entailed making a thick conserve from the tops of young plants sweetened with honey. This was spooned into young children on going to bed to alleviate nightmares. Nervous headaches disappeared under catnip's influence. Our grandmothers squeezed some of the juice from the green plant and took this in tablespoon doses three times a day. Back in the days when scarlet fever was a constant fear, the town health officer came and posted a quarantine notice on the door of a house where someone had contracted this infectious disease. Then catnip was mixed with equal parts of saffron, the delicate yellow stamen of the crocus flowers, and grandmother administered this concoction to the patient in a honeyed tea to break the fever.

Whatever you do, don't chew the roots. The old tale is told that chewing catnip roots turns the gentlest person into a fierce, quarrelsome ogre. Once there was a hangman who couldn't screw up his courage to do his nasty job until he had chewed a sufficient amount of catnip root.   [25]

THE CATTAIL

The cattail comes close to being a wonder plant. Cattails provide excellent wildlife and fish habitat. When flooded, they create a sheltered environment for aquatic insects and other invertebrates. They supply cover and shelter for bait-fish, spawning areas for sunfish and northern pike, and foraging areas for other game fish.

Marsh birds like yellow-headed blackbirds, red-winged blackbirds, sora rails, and marsh wrens would be homeless without cattails. Emergent stands of cattails furnish nest sites for canvasbacks and other diving ducks that nest over water, and they provide brood cover for these and many other ducks. Large stands of cattails serve as shelter for white-tail deer and make excellent winter cover for pheasants. The succulent stems and thick rootstocks of cattails are important foods for muskrats and beavers.

Human uses of cattails are almost as varied as wildlife uses. American Indians made floor mats and baskets from the leaves and diapers from the soft, downy seed heads; they found delectable ways of eating the starchy roots and protein-rich male spikes. Cattail leaves are still used in the manufacture of rush seats for furniture. According to Julian Steyermark's *Flora of Missouri*, the fluffy cattail seed heads have been substituted for cotton and wool and have been used as buoyancy stuffing in life jackets, as home insulation, and as soundproofing material. Cattails have been experimentally tried in alcohol production and paper and rope manufacturing. Euell Gibbons championed the cattail as the "Supermarket of the Swamps" and listed no fewer than a dozen ways of eating it, ranging from "Cattail Root Pancakes" to "Cattail Casserole."

Like all plants, cattails need oxygen to respire; they will drown if they are completely inundated for a couple of months. [26]

HERB IDEAS

Looking for something different for your garden? Have you tried herbs? Herbs have many uses, with the primary ones being fragrance and cooking. Lavender and rose petals are the herbs most commonly used for fragrance in sachets, pomanders, tussy-mussies, and bathwater, among many other uses. The most common use of herbs, however, is for cooking.

The herb most often found in gardens is chives. A low-growing plant (6–10 inches tall), it can be cut back repeatedly to a few inches high and still produce well. It also makes a nice pot plant for the kitchen windowsill. Noted for its mild onion flavor, this herb is often combined in cheese or seafood omelets, quiches, soups, stews, and sauces, or as a topping for salads and vegetable dishes.

Parsley is another very common herb in most gardens and is used primarily as a garnish. It can be used to flavor soups, stews, sauces, meats, and main dishes and in stuffings for fish and poultry.

Basil is probably the next most important herb in many gardens. A pungent herb, somewhat similar to licorice, it has many uses. Basil is frequently served with fresh or sun-dried tomatoes, cheese, and oil and vinegar as a salad. It is the key ingredient in pesto, a tangy combination of cheese, pine nuts, and olive oils used as a sauce for pasta and other dishes. It is tasty when used on fish, poultry, and most meat dishes. In addition to its culinary uses, basil is ornamental in the garden. With colors ranging from light green to dark purple, it goes well in the "edible landscape."

Another herb you may have or wish to grow is dill, most commonly known for its use with pickles. But dill is much more versatile than that. It's especially popular in Scandinavian dishes featuring meatballs, salmon, and other types of fish.

Marjoram is a sweet, spicy herb with a hint of mint. It complements the flavor of salads, vinegars, eggs, and casseroles. Oregano is famous for its use in pizza but adds a nice flavor to other Italian foods, fish, and poultry. Sage is known mainly for its use in breakfast sausage and poultry stuffings. Thyme's strong flavor makes it a popular herb for flavoring a wide variety of dishes. Different thyme varieties, such as anise and lemon, will add further dimension to your cooking. [27]

PREPARING FOR A SOIL TEST

Every few years I have my soil tested to check its fertility and pH. Adequate fertility ensures that necessary nutrients are in the soil, and proper pH makes it possible for plants to use those nutrients. It also provides a friendly environment for beneficial soil microorganisms.

Late summer and autumn are good times to sample soil. Soil-testing laboratories are usually not as busy as they are in late winter or spring, so you'll have the test results in hand in plenty of time to mix necessary amendments into the soil before spring planting. I have my soil tested at my state university's soil-testing laboratory (via the Cooperative Extension Service).

If your garden consists of two or perhaps more large areas where soil conditions obviously differ (if the soil changes from clay to sand, for example, or from well-drained loam to soggy muck), each of these areas will require a separate test.

Because laboratories use only a small amount of soil for the actual test—one cup, usually—the sample must be as representative as possible. To account for small differences in the test area, you will need to take at least a half-dozen samples from various spots within that area.

The roots of most vegetables feed in the top six inches of soil, so each sample must be six inches deep. Brush aside any manure, compost, or plant residues on the surface of the soil, dig a hole with a trowel, then slice a uniformly thick section of soil along one side of the hole. Carefully follow any packing instructions provided by the soil-testing laboratory. If you are having more than one area analyzed, label each sample before mailing it, and make a note to yourself of its location on your property. In a few weeks' time you should receive test results and specific recommendations for fertilizing and for altering or maintaining the pH.   [28]

WHITE BIRCHES: TWO VARIETIES

White birches, which contribute attractive yellows to fall's array of color, are easy to identify because of their white bark. Actually there are two different types of white bark: one that peels and one that does not. The peeling bark belongs to a species called white, or paper, birch. North American Indians used thick sections of this waterproof bark to build canoes. They stretched it tightly over a lightweight wooden frame, sewed the sections together, and caulked the seams with pitch. Designed for shallow waters, the birchbark canoe was the prototype for our modern canvas-covered canoe.

Though it is tempting to pull away sheets of this loose-hanging bark, stripping exposes the tree to disease organisms and leaves a permanent scar. Instead, look closely at the paper-thin layers and touch the bark to feel its chalky texture. The horizontal lines that look like dash marks are called lenticels. They are breathing pores that allow oxygen, carbon dioxide, and other gases to pass through the otherwise impermeable bark.

The other white-barked birch—the one with the bark that does not peel—is called gray birch, but it is "gray" only in comparison to the paper birch. Both white-barked birches have dark triangular markings where their branches join—or have fallen from—their trunks, but the gray birch's are more pronounced. They look like black mustaches.

Often gray birches are bent in graceful arcs. During the winter, heavy snow weighs them down until their tips touch the ground. When the snow melts, the trees spring back, but many remain partially bowed. Robert Frost immortalized this feature in his poem "Birches," where he describes a lonely country boy who climbs to the tops of slender gray birches and swings to the ground again. The poem closes with the famous line, "One could do worse than be a swinger of birches."

Although both these white-barked birches have yellow foliage, the leaves are quite different in shape.

The yellow birch is golden-barked and peels in tight little curls rather than in loose sheets. If you spot a yellow birch, break off a small twig and chew on it for a taste of wintergreen. The black birch grows mostly in southern Vermont and to the south. It has dark, nonpeeling bark, which makes it more difficult to spot and identify. The sweet sap of black birch is used to make a soft drink called birch beer.   [29]

AUTUMN'S HIDDEN HARVEST

The annual event is so obvious that most people do not comprehend its significance. To homeowners, it usually means getting out the rake and toiling for hours. But in actuality, the falling of leaves each autumn is one of the great transfers of energy on the face of the planet. Just ask a knowledgeable fisherman.

Leaves are critical to the perpetuation of life in a woodland stream. Some drop directly into the water, others are blown in. After immersion, they provide an abundance of organic carbon free for the taking. And generally, there are a lot of takers.

Organic carbon is found in anything that once was alive, be it a tree leaf, a turnip, or a T-bone steak. In fallen leaves, this carbon serves as the basic source of energy for a number of aquatic insects and crustaceans known as "shredders"— tiny creatures that literally shred leaves and other plant detritus. In turn, the shredders serve as savory fare for trout and other fish that anglers like to catch and eat themselves.

In the last twenty years, scientific understanding of stream life has undergone a revolution. Ecologists now realize that woodland streams are not merely conduits that quickly carry away leaves and other detritus from forests. In fact, just the opposite is true. Headwater streams, in particular, are very retentive and efficient in "spiraling"—the latest buzzword in stream ecology—materials such as leaves, and making them available to a wide variety of organisms as food. In a classic paper published, fittingly enough, in the autumn 1973 issue of *Ecological Monographs*, researchers Stuart G. Fisher and Gene E. Likens reported that Bear Brook, a mountain stream in the Hubbard Brook Experimental Forest in New Hampshire, derives more than 99 percent of its energy from leaves and other detritus from the land.

A number of stream ecologists believe that shredders derive their nutritional needs not so much from the leaves themselves but from the fungi and bacteria in a leaf.

Shredders eat leaves from different tree species, but, like gourmets, they have their favorite dishes. Leaves from the alder, basswood, green ash, and black ash rate four stars. Alder is particularly high in nitrogen, which is needed to make protein. By contrast, shredders seem to avoid oak leaves, which are tough and take as much as a year to become properly coated. [30]

SUMAC

The sumacs contribute their share to the autumnal pageant. They glow in scarlet and gold, often deepening to crimson and orange, and fling their magnificent beauty along fences, over deserted fields, and up rocky, gravelly mountainsides.

Most of us are acquainted with the staghorn sumac (so named because of the resemblance of its rich, velvety, thick, leafless twigs to the antlers of the stag when "in the velvet"), but we are more conscious of its presence when in winter its antlered branches are held aloft like candelabra, its pointed red fruit clusters silhouetted against the snow-covered landscape like flaming torches.

At this time of the year, when the tree—or shrub, if

you prefer, for it is not always a tree—stands naked, its architecture is revealed as being rather unshapely, somewhat stiff, awkward, and clumsy. In summer, when in full foliage, the tree acquires an entirely different, almost beautiful character, with fern-like leaves that lift and sway with every passing breeze. Among the leaves appear the whitish-green or yellow-green flowers in dense, conical, hairy clusters, with the staminate and pistillate on separate trees. Those of the latter develop into tiny, globular, acid drupes that are covered with deep red hairs and are clustered in large, compact panicles that remain on the tree during the winter and provide a festive board for birds as well as other forms of wildlife (in spite of the fact that the fruit is sour).

There are some 120 species of sumacs, 16 found in North America, and only 4 of them are trees. They form a temperate-zone genus of a great tropical family (Cashew) that contains some 400 species, which are widely distributed throughout Africa, Asia, North and South America, the Indian archipelago, Australia, and the Sandwich Islands. As for the name *sumac*, which can also be pronounced "shumac," it is said to have been derived from the Arabic name for a species occurring in the Mediterranean zone—*simmak* or *summaq*, the latter according to Webster. [31]

ALL ABOUT PUMPKINS

We have relegated the pumpkin to the role of Halloween jack-o'-lantern or pumpkin pie, but it was not always so. And it deserves a better fate!

Pumpkins were an integral part of early American culture. They were used to such an extent by the early settlers that an unknown writer in 1620 wrote, "We have pumpkins at morning and pumpkins at noon. If it were not for pumpkins, we should be undone."

The word pumpkin comes from an old French word,

"pompion," which means "cooked by the sun." It was native to tropical America and was first used by the Indians. But pumpkins were not limited to America. In China it was called "Emperor of the Garden," and when introduced to Europe, it was called "Turkish Cucumber" because it came by way of Turkey.

Pumpkins were often cut into cubes and strung up to dry for later use. Stewed pumpkin was mixed with Indian meal for bread. It was used as a vegetable, either mashed or cubed. Hollowed-out pumpkin filled with milk and baked made a simple pudding.

Nutritionally it rates high. One-half cup of pumpkin provides more than the recommended daily allowance of Vitamin A. It contains a small amount of iron, as well as other vitamins and minerals. One cup contains only seventy-five calories—considerably less than winter squash or sweet potatoes. When used in pudding with milk, it is higher in nutrition. You might like to try pumpkin as a vegetable. Peel and cube it and cook until just tender. It's delicious flavored with butter or margarine and salt and pepper. Or you can mash it like squash. Use the small pie pumpkins rather than the large jack-o'-lantern variety—they're sweeter and more flavorful.

The easiest way to cook a pumpkin for the pulp is to cut it in half crosswise, clean out the seeds, put it back together, and bake it in a slow oven (300°–325°) until soft. Then put the pulp through a food mill or strainer.   [32]

WHEN PLANTS MARK TIME

Springtime's green-up doesn't just happen; nature has been preparing for this event for months. The red oak acorn beneath its coverlet of dead leaves where it rolled last autumn began as a single, microscopic cell that was fertilized by sperm from a pollen grain last May. Then began the almost interminable dividing into more cells—the process by which it became an acorn and by which, with luck, it will eventually become an eighty-foot giant.

But now, as winter progresses, it merely bides its time. Like other seeds—from the dustlike seeds of the orchids to the robust black walnut—the acorn contains an embryonic plant.

With the coming of spring, most seeds absorb additional moisture and split their shells. A simple sproutlike extension grows upward, often surmounted by the cotyledons, or seed leaves, which contain the stores of food for the burgeoning embryo until its new roots and foliage become functional.

Many mature plants, perennial herbs especially, survive the winter only as underground structures—roots, rhizomes, tubers, or bulbs. In addition to their obvious function of anchoring and supporting the stems and leaves while alive, these structures also contribute to the spread of the species. Bulbs separate and sprout. New plants spring from trailing rootstocks. Tubers, widely separated from the parent, send up shoots from their "eyes."

But more important, these buried parts are veritable wintertime pantries. All summer long the leaves manufactured sugar and stored it as starch and sugar in the roots, tubers, rhizomes, or bulbs; and this food will nourish the new plant when it emerges in the spring.

Some young plants, usually biennials, which require two growing seasons to produce seeds, have another way of getting the jump on their neighbors. They do this by forming so-called winter rosettes in the fall—looking for all the world as though some furtive gnome has grabbed hold of a normal plant's roots and pulled it down to the ground, leaving its foliage radiating from a common center on the surface. The purpose of this ground-hugging arrangement is to protect the plant from the rigors of winter, and plants thus compressed usually remain green until spring. When warm weather arrives, a lengthening of the stem makes it a full-sized plant.

Does the method work? We must assume it does, for its most conspicuous exponents—the thistles, evening primrose, chicory, mullein, wintercress, the goldenrods, and the mustards, to name a few—are among our most indomitable weeds. It appears that plants, in common with nearly all wildlife, must get the better of winter to survive.   [33]

WATERING HOUSEPLANTS

Of all the things done to houseplants, watering is probably the one cultural practice that causes the most problems. Watering is affected by the type of soil used, light exposure, temperature, and humidity. Many factors determine when a plant needs to be watered. Just as many environmental factors change with the seasons, so do the water needs of plants.

The two main aspects of watering to be considered are frequency of watering and amount of water applied. The watering frequency is simply how much time passes between waterings. The frequency will vary over the course of the year.

Avoid watering on a fixed schedule such as every week or every five days. A fixed schedule does not necessarily give plants water when they need it. In fact, watering on a fixed schedule may mean plants are overwatered at one time of the year but underwatered at others. With few exceptions, plants should be watered when the soil feels dry to the touch. This means the frequency of watering will vary with the rate at which the soil dries out.

Apply enough water so some comes out the drain hole at the bottom of the pot. This flushes out salts which can lead to root injury. Do not let plants sit in excess water. It will be reabsorbed, and the salts dissolved in the water will thus be reabsorbed.

There is no easy way to tell when plants in undrained containers have had enough water, but the actual technique for watering is not difficult. The best way is to use a watering can with a long, narrow spout. This allows the placement of water onto the soil.

Watering plants with dense foliage will be more difficult

if this type of watering can is not used. Try not to put the water on the leaves and crowns, as rot diseases are more likely to occur if water is continually poured on the crown. For watering, tepid water or water near room temperature is best.

Bottom watering is a practice where the plant is set in and absorbs water from a container filled with water. Plants regularly watered from the bottom should occasionally be watered from the top to get rid of excess salts in the soil. The bottom layer of soil becomes saturated when a plant continuously sits in water. Any roots growing in this saturated layer will die. The soil available for healthy plant growth is thus reduced. [34]

CACTI

Want a good exercise for your green thumb this winter? Grow cacti as houseplants. True cacti, a type of succulent, are native to the Americas. Many types may be grown as houseplants, each different in size, color, shape, and flowering habit. Among the most popular types are the star cactus, golden barrel, old man, bishop's cap, bunny ears, rattail, pincushion, turk's cap, and ball cactus.

Most cacti purchased at plant shops, garden centers, florists, and grocery and discount stores are grown alike. They have one thing in common—they all prefer a growing space with plenty of sunlight. Cacti grown on windowsills facing south usually flourish. The next best exposure is the light from an east or west window, since it can provide direct sun part of the day. You can place the plants outdoors in summer. Many gardeners think there is too little light in our area during the winter, but this is often not the case. With snow cover, more light is reflected during the winter than penetrates through shade trees around the home in summer.

Perhaps the main trick to growing cacti is proper watering. Many cacti have been killed from overwatering during winter—including mine! If the weather is cloudy or even predicted to be cloudy, don't water. If in doubt whether the soil is dry, don't water. When watering, apply only a small amount to moisten the soil area around the roots. Allow the soil to become dry before additional watering.

Maintain the temperature during the growth period (usually spring and summer) at 60°F at night and 10–15° warmer during the day. During the dormant period (usually fall and winter), reduce the temperature to 45–55°F. If most of your rooms are warmer than this, then place your cactus near a window (but not touching it), where the temperature may be 5–10° cooler than in the middle of the room.

Fertilize cacti several times during the growth period with a liquid fertilizer. Products labeled 5–10–5 or 10–20–10 and those containing fish emulsion are all suitable. Follow directions on the label carefully.

For repotting, use a soil mix prepared for cacti or make your own from one part coarse builder's sand, one part loam soil, and one part peat moss.

Finally, if you brush against your cactus and get a few small spines stuck in your fingers, use heavy-duty tape (sticky side to the spines) to pull them out! [35]

SPECIALIST OFFERS TIPS ON BUYING FUELWOOD

There's more to buying wood for home heating than simply phoning a supplier and asking when he or she can deliver. Since fuelwood may be sold by weight, load, or stack, it's important to know just how much wood you are getting before you agree on a price. "The heat value of an air-dried standard cord of hardwood, when burned in an efficient wood-burning unit, is equivalent to 130 gallons of No. 2 fuel oil," says Dan Bosquet, extension forest products specialist at the University of Vermont. "So if you choose to heat with wood, you can generally save money. The catch is that the actual amount of wood in a cord may vary according to how the wood is stacked, the size and straightness of the sticks, and whether they are whole or split. A cord containing greenwood and softwoods will not, of course, burn as efficiently as dry, seasoned hardwood. Although it may be cheaper, it is not really a bargain."

A standard cord of wood is defined by the weights and measures law as a pile 4 feet high and 8 feet long, made up of sticks 4 feet in length. It contains 128 cubic feet of both wood and air. A face cord is a pile 4 feet high and 8 feet long consisting of pieces of wood of any length. It also is called a short cord, a run, a rick, or a rack, though regardless of the name, it contains less volume of wood than a standard cord.

"When buying fuelwood, also keep in mind that different species of wood have different burning characteristics. How much heat is obtained from the combustion of different kinds of wood depends on the concentration of woody materials, resins, ash, and water. In general, the heaviest woods—such as hickory, oak, and ironwood—when well seasoned, have the greatest heating value per cord. Lighter woods—including aspen, basswood, and willow—provide the same amount of heat per pound of wood but give less heat per cord because they are less dense. Elm is not a good choice, as it is difficult to cut and has a high moisture content and lower heating value than most other woods."

For good burning, wood must be allowed to dry for six to nine months after cutting. Splitting helps reduce drying time and makes handling, piling, and burning easier. Stacking allows for proper drying and should be done right after splitting. Cover the top of your woodpile with a waterproof tarp, allowing for good air circulation throughout the pile.

[36]

GENETIC IMPROVEMENT
OF SUGAR MAPLE

Our maple orchard, or sugar bush, as New Englanders call their maple tree groves, is a delight to walk through any time of the year—especially in late spring with wildflowers underfoot, or in the brilliant fall. But best of times to the maple sugar maker is February, when he sets out to tap his trees. Sugaring, like everything else the farmer does, is a risky business. The weather can, as always, be either too warm or too cold, and no one is ever sure of how many gallons he'll make. We knew that certain trees gave more sap than others and suspected that some sap had a higher sugar content. Now a friend of ours, Howard Kriebel of Ohio State University, a plant geneticist, has sent us several papers he has published on his studies on the breeding of sweeter sugar maples.

He cautions that applying biotechnology to forestry is not as easy as applying it to plants. Tree crops take decades to develop, and gene recombinations by breeding over several generations may take a hundred years or more.

Ohio is a maple-sugar-producing state, as are Maine, Vermont, New Hampshire, New York, and the Province of Quebec. But Ohio has been interested for many years in bettering the quality of its trees. Genetic research started as early as 1955 in Ohio with testing of local trees by taking a few drops of sap from a small puncture made in the trunk at tapping level. Cuttings were taken from the best of these trees in the late winter of 1957 and grafted in a greenhouse onto randomly selected rootstock potted the previous year. In 1958 the grafted trees were planted to produce seeds. The saplings grown from the seeds were then distributed to Ohio landowners for planting.

It is now thirty-some years since the initial testing was done, and the first sugar trees developed from that testing will come into production around the year 2000. The saying among sugar makers is that it takes about 40 gallons of sap to make a gallon of syrup. But the prediction is that these genetically improved trees will produce a gallon of syrup from 18½ gallons of sap. What will this mean for the future? Will our magnificent old giants have to be cut down to make room for these maple tree babies?   [37]

WHY DOES THE SAP RUN?

Although maple sugaring is an annual season in the same sense that winter and spring are annual seasons, it doesn't begin with a solstice or an equinox or even a specific date on the calendar. It begins sometime in late February or early March when the weather achieves a perilous balance between winter and spring. When it's cold at night and warm during the day, trees begin to activate for spring. That's when we tap maple trees to borrow some of their sweet sap for our syrup-making operations.

The big mystery to sugar makers and scientists alike is what makes sap do what it does at this time of year—especially what causes the sap to run on warm, sunny days. The current theory, and there have been many, is that when the weather warms up to above 32°F, maple sap expels carbon dioxide. This carbon dioxide forms bubbles, and these bubbles collect in the fibers that surround the vessels that conduct the sap. The expanding carbon dioxide squeezes the vessels and puts pressure on the sap in the vessels. This pressure drives the sap upward and downward toward the points of least resistance.

Sap moves up and down in the sap-conducting vessels, but there is no significant lateral movement inside the tree trunk. In fact, two tapholes right next to each other are completely independent of one another because each has severed a different group of vessels. By severing the vessels, the taphole creates an unexpected point of low resistance, and some of the sap that would have been pushed upward or downward by the pressure of expanding carbon dioxide is diverted right out the sugarer's spout into his waiting sap bucket.

If the weather stays warm for a while, sap stops running. Without a return of cold weather, the taphole would run dry and stay dry. Cold weather allows the tree to recharge itself. The carbon dioxide contracts and reenters the sap solution. More water is drawn up into the sap-conducting vessels through the root system. When it warms up again, the carbon dioxide expands, exerts pressure on the sap, and more sap moves up and down the tree and out the tapholes. All this is still only theory, however, although it's based on controlled scientific experiments and observations.

We become increasingly sophisticated in methods of tapping, collecting, and boiling sap and increasingly sophisticated in our understanding of tree physiology, but the only important thing is that maple trees are constant. We are fortunate to have the weather conditions that invite our maple trees to perform so readily and deliciously each year at this time, as winter gives way to spring.   [38]

ANIMALS

KEEPING WARM:
FIVE-DOG NIGHT

Richard Porter of East Charleston, Vermont, aged seventy-five or eighty, does not own an electric blanket. He does not have central heat in his three-room cabin, has not heard of modern airtight wood stoves, does not own a kerosene or gas space heater, and regularly allows the fire in his box-type wood stove to burn itself out each night around eleven o'clock. He is not averse to cold drafts and for this reason has never insulated the pine board walls of his cabin, even though the temperatures in East Charleston commonly dip below zero degrees Fahrenheit for weeks at a time. And yet, in spite of his apparent lack of conveniences, Porter says he is never cold at night. He has devised a system of living blankets which automatically pile themselves on his bed in response to the temperature.

Like many who have deserted human society, Porter keeps a number of dogs for companions. Not surprisingly, it is they who keep him warm at night. Each winter night, about the time the box stove begins to cool, the first of Porter's alternative heating systems—a black and tan hound named Spike—begins to stir from his spot beneath the stove. Spike will climb into Porter's bed when the room temperature reaches 50°. Louise will get up around 40°. Any colder and the others begin to come in through a dog door which Porter has cut in one of his door panels.

Spike and Louise, his favorites, spend most of their time in the cabin. The others come in only to sleep and only when it's cold. They come in a progression, Porter says. Jeff, a collie-like dog with a thick coat, will move in on those nights when the outside temperature reaches 10° and will join the others on the bed shortly thereafter. Alice, a medium-sized dog of indeterminate parentage, arrives after the temperature dips below 10°. But those nights when the mercury dips below zero mark the arrival of the warmest dog of all, an immense golden-eyed thing named Bull, who has a strong shot of Irish wolfhound in his blood.

Porter says that Bull does not normally appreciate such bourgeois comforts as warm stoves and human companionship. But in his aloof, dog-like way, he is as devoted to Porter as any dog of his type could be. Porter believes that it is generally below Bull to come in at night, let alone climb up on the bed with the lesser beings in the pack. But zero-degree nights get the better of his pride, and invariably he deserts his usual hideout beneath the porch stairs and squeezes in through the narrow door panel. With Bull on the bed, there is not a night that Porter cannot endure.   [1]

MESSAGES IN THE SNOW

Gaze upon freshly fallen snow and the winter landscape appears quiet and peaceful. Nature seems to be on hold, the animal kingdom fast asleep.

But on close examination, the blanket of snow tells a different story. For animals that neither migrate nor hibernate, winter is a time to hustle for food and shelter. It is no time to stand still. A covering of snow becomes a text filled with anecdotes and messages—many, but not all, written clearly in the language of animal tracks.

Only a little sleuthing is needed to reconstruct events that occurred earlier in the snow. But to read the story clearly, it is best that snow be wet and no more than three inches deep. Also know that there are three basic clues to look for when identifying tracks: the habitat in which a particular track is found, the actual track pattern, and the presence of any auxiliary signs.

The gray squirrel lives in deciduous trees, particularly those that bear nuts. If you see small, "four-print" tracks (a group of four footprints followed by another group of four) leading to or from such a tree, then the habitat clue suggests squirrel prints. Cracked nut shells nearby are good auxiliary evidence, as is a leaf nest within the tree.

Red foxes leave a doglike paw print in a straight line, usually across open fields where the animals prefer to hunt. Each print is rounded and may have impressions from foot pads and toenails. Auxiliary signs left by red foxes include urine marks and doglike droppings.

The four-print track of a hopping cottontail looks as if it were put together backwards. The two side-by-side marks from the rabbit's large hind feet come first, showing where the animal landed. Just behind come two smaller front foot marks, indicating where the rabbit pushed off. Gnaw marks on the trunks of shrubs or on trees in brushy fields provide another clue.

The white-tailed deer leaves a sharp-pointed, heart-shaped track. The two large front pads of the hoof may be followed by two dewclaw marks in the rear of each footprint. Most deer drag their feet, particularly in deep snow.

For people who live south of the snow belt, similar clues can be read in mud or sand. But snow is the best base for prints. And that makes winter an excellent season for reading the daily wildlife news written in tracks. [2]

COMMUNICATION

Danger is signaled in animals in a variety of ways. The white-tailed deer raises its highly visible white tail as a warning. The beaver's tail slaps the water with a loud smack. The smell exuded by wounded fish announces alarm with amazing speed to fish of the same species. The response to such alarms is immediate flight. The absence of sound causes concern among certain voles which emit an ultrasonic squeak as they feed amongst dry grasses and leaves. This squeak inhibits dispersal behavior. A nonsqueaking leaf rustler causes instant scurrying to safety.

Many communication devices are ritualized behaviors which have grown out of the normal process of living. Birds about to take flight crouch, raise their tails and spread their wings. For some, this behavior has become ritualized, and color markings on wings and/or tail, when exposed by this posture, serve as a signal to members of the same species to prepare for takeoff. The male of one species of dancing fly wins the female's assent by presenting her with "an empty silk balloon." Research into behavior of related species shows that this evolved from the male's giving food to the female so she wouldn't eat him before he had the chance to identify himself. To make the edible gift more conspicuous, it was wrapped in silk. Over time, the food was forgotten and the symbolic package now suffices.

Successful communication depends on one or more of the senses. Pheromones, chemicals released by creatures to stimulate one or more specific reactions, involve the senses of smell and/or taste. A harvester ant responds to a small amount of a certain alarm pheromone by moving toward the odor to check its source and, if possible, to defend the nest. A greater amount stirs the group to an alarm frenzy and, if that amount persists for more than a minute or two, an instinct to dig, probably to escape, takes over. Barnacles, clinging to the rocks of tidal waters, give off a pheromone that invites barnacle larvae to settle there and mature.   [3]

CAMOUFLAGE

Sometimes camouflage hunting clothes come in handy. There are outfits that make a person look like a pile of leaves, tree bark, or a bunch of cattails. A lot of animals have been using camouflage for a long time. And from them man has discovered how effective it can be.

Some animals hide with the help of camouflage so they *won't* get eaten. And some animals use it to hide so what they *want* to eat will get close enough to catch. A very important thing for an animal that is sought for a meal is to be unnoticed. But that doesn't always mean it has to be camouflaged to look like a stick or clump of grass to throw off the hunters. Sometimes all that is necessary is to hang around with something a lot more colorful than it is.

In birds, the males are often brightly feathered. That can get them a lot of attention from predators. And to top it off, they usually make a lot of noise. The male, therefore, is more likely to be eaten than the plainer, quieter female. Actually this seems to fit into nature's plan, because the female lays the eggs and is more important.

There's an old hunter's trick to look for an eye when searching for hiding game. And nature has paid special attention to a lot of eyes to make them hard to find. In some animals, the colored part of the eye matches the skin around the eye. And sometimes spots or stripes on the head of the hunted animal keep the hunter from easily discovering an eye. Then there are animals with broad, dark bands across their eyes, such as the "masked" raccoons.

There are also creatures that use fake eyes to help protect them. Since an eye makes such a good target to a hunter, many butterflies have developed dots on the edges of their wings. Birds will stab these "bull's-eyes," and instead of getting in a killing blow to a body will peck the edge of a wing.

But even though fake eyes are popular in some circles, the name of the game is usually not to draw attention, either as the hunted or a hunter. A shadow can be a dangerous thing. It can hide an enemy, or a shadow can give away a hunted animal. To keep down the danger of a shadow giving it away, a rabbit crouches through the day in a low spot called a form. Young birds following their mothers through the woods will squat down when their mothers give the danger signal. This keeps their shadows from giving them away. Even butterflies are very careful about their shadows. When they are at rest with their wings folded up over their backs, they like to keep turned facing the sun so their wings make only a thin line of a shadow instead of the broad patch that would be there if they were sideways to the sun.   [4]

COLOR IT SURVIVAL

The variety of living colors that we marvel at in nature exists for more practical reasons than mere beauty. There is evidence that the purpose for every shade and pattern displayed by animals and plants is simply survival. But this single purpose is served in many ways.

For some creatures, color provides camouflage. The ground-nesting American woodcock blends remarkably well with the dead leaves upon which it incubates its three to four eggs. The eggs and, later, the hatchlings themselves are concealed by their color. Likewise, the spotted coat on a white-tailed deer fawn helps hide the creature from predators by breaking up its shape in the sun-dappled woodlands where it lives. When danger is near, the fawn lies hidden while the doe exposes herself to create a major distraction.

The American bittern's long, striped neck "disappears" when held parallel to the cattails and grasses in which it lives. Sometimes the bird will even sway with the grasses in a breeze, further camouflaging itself from both shoreline predators and from the fish that it preys upon.

Countershading—differences in color between an animal's belly and back—can also provide important protection. When bright sunlight reflects off the dark back feathers of the lesser yellowlegs and casts a dark shadow on its light-colored breast, the bird's form very often "melts" into its shoreline habitat.

In contrast, some plants and animals *want* to be seen.

The bright red petals of the cardinal flower attract hungry insects and ruby-throated hummingbirds. While feeding, these winged creatures are dusted with the flower's pollen, which they carry to other cardinal flowers. In the process, they unwittingly cross-fertilize the plants. Meanwhile, with some animals, color plays an important role in signaling the abundance of food. Masses of white gulls and black vultures, for example, can frequently be seen many miles away by others of their kind—a silent announcement of sorts that a feast is in the offing.

Some frogs and insects advertise their presence with conspicuous—sometimes gaudy—colors to warn predators that they are foul-tasting or even poisonous. Such creatures are so successful in avoiding predation that the process of evolution has mimicked their colors and patterns in edible look-alike species. [5]

A LOOK AT ANIMAL VISION

Imagine that you are standing in a green, grassy meadow filled with colorful wildflowers. A plump, grayish-brown rabbit scampers from one plant to another, stopping frequently to survey the area, scarcely disturbing the meadow's peaceful silence. Does the rabbit see what you see when it looks at the field?

The answer is no. A person standing in the meadow sees green leaves and pink, red, purple, and yellow flowers. To the rabbit all objects appear black, white, or gray; it sees the shapes we see but can't detect the colors. Yet the rabbit has no trouble making its way through the field, distinguishing one plant from the next, and watching for enemies.

In fact, rabbits can see more of the meadow than humans can. Because our eyes are in the front of our heads, we see directly ahead of us but only partway around us. A rabbit's eyes bulge outward and are set far back on the sides of its head. As a result, it can see almost in a complete circle without moving its head or body.

This wide field of vision is more important to a defenseless rabbit than the ability to see colors. It enables the rabbit to see bobcats or other predators approaching from any direction without drawing attention to itself. With luck, this gives it enough time to escape. In contrast, the bobcat has eyes set in the front of its head, just as we do. It doesn't get an all-around view, but it has excellent vision for hunting.

Animals rely on vision to find food, escape from enemies, and find and attract mates, and every animal's eyes have adapted to meet its specific needs and environment.

When we see an object, we actually see the light that the object reflects. Sunlight is made of many types of light, and each type produces a different color when it hits an object. An object's chemical structure determines which colors it will absorb and which colors it will reflect. A leaf, for example, reflects green light. Our eyes receive the reflected light and change its energy into a form that our brains understand as a picture.

Eyes have millions of tiny nerve cells (called photoreceptors) that recognize light or patterns of light. There are two types of photoreceptors: rods allow eyes to see in dim light, such as night light, while cones function in bright light, such as daylight. Rods and cones are found at the back of the eyeball, in the retina. The retinas of humans, birds, and many other animals contain more cones than rods; this allows them to see colors and sharp detail. [6]

THE BRUMATION OF ECTOTHERMS

What do you do if you're a cold-blooded vertebrate when the cold of winter sets in? What do you do when there's three feet of snow, and the temperature is well below zero, and the only thing you've got for protection is a set of scales or a bunch of glands in your skin? You can't run very fast if you're a turtle, or very far, and if you linger in the warm autumn sun one afternoon, and the next day the temperature drops thirty degrees, your body movements slow like molasses.

Not long ago it was thought that amphibians and reptiles disappeared into the muck of ponds and lakes or crawled ever deeper into the bowels of the earth to escape the consequences of winter. There they would endure the cold season in a sort of suspended animation until the warm rays of spring encouraged them, somehow, to return to life. Most publications, including college texts, will relate this or similar scenarios, with some minor embellishment. But any kid who has ever sprawled out on the clear ice of a pond to watch the comings and goings of life in the water below knows a different story. I remember how startled I was the first time I saw an adult painted turtle cruise past my nose beneath the ice.

The truth of the matter is that winter activity for amphibians and reptiles is a bit different than is conventionally believed, as recent research is beginning to show. It turns out that some species of amphibians and reptiles have several options when it comes to winter survival strategies.

Some salamanders prefer cold, rushing streams; others thrive only in warm, shallow lakes; one never goes to water; another never leaves. There is a toad that lives only in sandy soils and a frog that rarely climbs down out of trees. There is a salt marsh turtle, an oak forest turtle, and another that likes only wet meadows. Some snakes undergo impressive migrations to locate historic winter den sites where dozens of individuals congregate; others wander singly, perhaps to find a crack in a house foundation and a new sanctuary for the season. There is a pronounced autumn movement of individual amphibians and reptiles that brings them to over-wintering sites.

Moving, normally overland, to reach overwintering dens or other locations, exposes individuals to several dangers. For one thing, it takes a fair amount of energy to move, and animals (even hibernating ones) need as much stored energy as possible to take them through the winter. Secondly, migration exposes individuals to predators and other hazards (such as moving automobiles) when they are relatively unfamiliar with the terrain. For some reptiles, particularly snakes, sudden cold snaps may initiate immediate, long-distance trips that take them several miles to their denning areas, if they make it. [7]

SNAKES MAXIMIZE SUCCESS WITH MINIMAL EQUIPMENT

Biologists such as Harvard's E. O. Wilson believe that when humans evolved in Afroasia in the midst of rich, venomous snake fauna, we developed an instinctive reaction to snakes. As Wilson put it, "The brain appears to have kept its old capacities, its channeled quickness. We stay alert and alive in the vanished forests of the world." Snakes, whether they are venomous or not, can all startle us when we are suddenly confronted by them because snakes are highly stereotypic in form. Their basic shape does not vary like those of other animal groups such as mammals, fish, and birds. Snakes are all stripped-down, streamlined predators.

Alexander Skutch, one of the pioneering tropical naturalists, captured the emotional impact of the economical serpentine form: "The serpent is stark predation, the predatory existence in its baldest, least-mitigated form. It might be characterised as an elongated, distensible stomach, with the minimum of accessories needed to fill and propagate this maw—not even teeth that can tear its food. It crams itself with animal life that is often warm and vibrant, to prolong an existence in which we can detect no joy and emotion."

Many people share Skutch's horror, but biologists find much to admire in what snakes have accomplished as predators working with limited tools. This homogeneous group, lacking limbs for prey capture and handling, manages to achieve one of the most varied diets of any group of predators. There are snakes that eat only slugs, that specialize in scorpions, that subsist on frog eggs. Some feed only on birds, others only on fish. To accomplish this diversity, snakes have had to make the most of their mouths.

Ancestral snakes were probably lizardlike and used a powerful bite to kill prey the way some large monitor lizards do. Burrowing snakes rely on biting with a little constriction to kill their prey. Boas rely far more on constriction and use the bite more to grab than to kill their prey. The largest pythons weigh well over 320 pounds, yet the heaviest prey they can eat weigh less than half of that. In contrast, a snake such as a pit viper can kill and ingest a food item even heavier than itself. There is a record of a fer-de-lance eating a lizard one and a half times its own weight, perhaps the ultimate feat of terrestrial gluttony. [8]

NIGHT FLYERS

Bats are mammals that bear their young live and then suckle them. Bats are not birds. They are not mice with wings. They belong in their own order, Chiroptera (which means hand wing), and are more closely related to shrews than to rodents.

Bats do not get into people's hair. I am not even sure how such an idea got started. The truth is that bats are among the least aggressive of mammals. A story is told of Konrad Lorenz, the famous animal behaviorist, who one day visited bat researcher Donald Griffin. Griffin had a vampire bat in a cage in his laboratory. Lorenz wanted to know what it would feel like to be bitten by a vampire. However, despite repeated pokings and proddings, the little creature could not be goaded into biting him. It simply would not attack. But like any wild animal, bats should not be harassed or handled.

Bats are very important in the ecology of an area in which they live. In a given evening a bat consumes enough insects to amount to 30 percent of its body weight. This means that an individual bat will eat four to eight pounds of insects per year. A large colony of bats, which might consist of one thousand individuals, will consume two to four tons of insects a year, thus providing a very effective control.

Bats can be induced to roost in an area by erecting bat boxes. These are similar to large birdhouses, but instead of a hole in the front, they have an open bottom with partitions about three-quarters to one and one-half inches apart.

To illustrate the bat's value in controlling insects, studies done in the early 1950s on the little brown bat showed that in the wild they can consume on average about five hundred mosquitoes per hour, or one every seven seconds. Whenever you see a bat make what seems to be a somersault in flight, it has netted a bug in its wings, transferred it to its tail membrane and bent over to take the bug in its mouth. All this quite literally in the wink of an eye.   [9]

EARTHWORM

The lowly earthworm doesn't seem to mean much to most people, yet it is considered by many to be our most useful animal. Many years ago Charles Darwin, in writing about earthworms (*The Formation of Vegetable Mold Through the Action of Worms*), said that "it may be doubted if there are any other animals which have played such an important part in the history of the world as these lowly organized creatures."

Many of us regard the earthworm as being useful only as bait for catching fish. But because of its underground activities, it is of inestimable value to the farmer and to those of us who have a garden, for it works unseen day and night, harrowing and fertilizing the soil for our benefit. It burrows twelve to eighteen inches into the ground and brings the subsoil to the surface; it grinds the subsoil in its gizzard and changes it into soil by secreting lime that neutralizes the acids in it.

The earthworm feeds principally on pieces of leaves and other plant material, particles of animal matter, and debris of various kinds that lie on the surface of the ground. Holding fast in its burrow by its tail end, it issues forth at night, for it is a nocturnal animal, and stretching itself to great lengths, it reaches out in all directions and grasps bits of food, which it pulls beneath the surface. Here part of it is eaten and the remainder passes into vegetable mold, which is so essential to the growth of plants. Furthermore, it enriches the soil by burying the bones of dead animals, shells, and the like, which, upon decaying, furnish the essential minerals for plant life. It even provides drainage by boring holes to carry off the surplus water, and by doing so, it promotes aeration. Indeed, the earthworm is not only a tiller of the soil, but it is also an agriculturalist as well, for it plants fallen seeds by covering them with soil, and it cares for the growing plants by cultivating the soil around the roots. As a matter of fact, if it were not for the earthworm, vegetation would not be so luxuriant, and much of the earth's soil would be useless to many plants.

The changing character of the landscape and much of the beauty of our fields and forests can be attributed to the labors of this diminutive workman; the earthworm can also be credited with having preserved many ancient ruins and works of art by covering them with earth. The familiar mounds of black earth, or castings, that we so often see on the ground or on our lawns are particles of soil swallowed by the earthworms in their burrows and brought to the surface. As there may be as many as 50,000 worms in an acre of ground, Darwin estimated that more than 18 tons of earthy castings may be carried to the surface in a single year, and in twenty years a layer three inches thick would be transferred from the subsoil to the surface.

Earthworms are strictly nocturnal animals and are not found outside their burrows during the day unless "drowned out" by a heavy rain. During the daylight hours they remain stretched out in their burrows with their heads—or rather, anterior ends, for they do not have heads in the strictest meaning of the word—near the surface. Apparently earthworms cannot find their way back to their burrows if they leave them; hence they anchor themselves to the walls as they extend themselves over the surface of the ground in search of food. Everyone at some time has seen a robin tugging away at a protesting worm in an effort to dislodge it; if you have ever tried it, you doubtless found it was not easy to do.

Run your fingers over the earthworm's body, and you will find it rough to the touch. This roughness is due to stiff bristles which project from the body and which you can easily see with a hand lens or magnifying glass. These bristles, or setae, protrude from small sacs in the body wall and can be extended or retracted by special muscles. When the earthworm wants to remain fixed in its burrow, it simply extends the setae out beyond the surface of the body and into the sides of the burrow, and it will be securely anchored—or relatively so. Then, when the animal wants to change its position, it retracts the setae and is free to move.   [10]

MINK

Even those who knew where the mink lived seldom saw her. A quick flash of brown, almost black in the dimming light of evening, was the most that one could expect when watching for the mink. Like most mammals, she reversed the human schedule, preferring night to day for trips outside the home.

The mink was known as "she" because male mink lack the tenacity to one location that this mink held. Female mink enjoy a home with a river for a front lawn and perhaps a sandbar for a backyard. Male mink are travelers who have a district that may cover several miles. Males have been known to travel fifteen miles in a night.

If one were to guess, one would say that there are more mink today than there were fifty years ago. The mink population, however, probably is hard pressed. The animals give much attention to their coats; therefore they do not do well in polluted water. The invisible chemicals that invade rivers also work against the mink's future. The chemical soup known as PCBs exists in parts of southern New England, and it causes not only stillborn kits but also death among adults. PCBs contaminate fish, which mink eat.

Mink will have their young in late April or, more commonly, in the first week of May. The kits are about the size of a human little finger. Usually three to four kits arrive. The young are weaned when about six weeks old but remain with their mother until late August, when the family breaks up. Mink are antisocial. Except when the young are being reared, the rule is one mink to one borrow.

Young mink start out life in a white coat. By the time the average young mink leaves the burrow, it will be wearing brownish fur. As members of the weasel tribe, mink belong to a group of animals capable of responding to season changes by switching from the browns of summer to whites of winter. Mink themselves do not experience seasonal color changes; however, the ability of cousins to do so may account for the variability of mink color forms in captivity.

[11]

THE GRAY SQUIRREL

Needle-sharp claws on its feet help the squirrel grasp tree branches and twigs and scamper across all but the smoothest surfaces. The squirrel's bushy tail, which stretches as long as its body, is a rudder when the animal leaps from branch to branch. Bulging, side-mounted eyes and sharp vision rule out sneak attacks on squirrels, as many sportsmen can attest. So does the squirrel's sprinting ability—twelve miles per hour over short distances.

Add to these physical attributes an apparent abundance of "smarts," plus a voracious appetite, and the gray squirrel has the makings of a four-footed paradox. For caretakers of parks and for people who feed birds, the animal is an often annoying, aggressive adversary that destroys plants and hogs seeds. But for admirers of backyard acrobatics, the creature is a delightful entertainer, conducting wild chases and romps through treetops, across lawns, over fences and porch railings—and, yes, straight onto bird feeders.

Aside from an occasional flea, notes gray squirrel authority Vagn Flyger, squirrels pose no serious threat to human health. They are overrated as carriers of rabies, he contends, explaining that because of the gap between the squirrel's incisors and cheek teeth, its bite does not contain saliva—a prerequisite to the spread of the disease. "Besides," says Flyger, "a squirrel bitten by a rabid fox, or any fox, isn't going to live long enough to spread anything."

Foxes are but one of the many gray squirrel enemies. Bobcats, hawks, owls, black rat snakes, and a host of other predators feed on them. A popular game species, the gray squirrel has been little affected by hunting throughout its range.

Flyger has learned that gray squirrels themselves eat a great deal more than hickory nuts, beechnuts, and acorns. Fruit, tree buds, garden corn, and—of course—birdseed are all fair game. Insects and meat, including an occasional baby bird, make up about 10 percent of their diet. Squirrels also eat significant numbers of mushrooms, including the deadly amanita and others fatal to humans. The reason for their immunity is not known.

Probably the most-asked question regarding gray squirrels is whether they really find all those nuts they bury in the fall. The answer, according to Flyger, is a qualified "yes." About 95 percent of buried nuts are found—but often not by the same animal that squirreled the nut away in the first place. The secret in nut finding is not memory, but an acute sense of smell.   [12]

A LOOK AT HOUSE MICE

House mice have been associated with human beings since the early days of agriculture. The ancestral species inhabited the dry grasslands of central Asia, where they subsisted on seeds and insects. When human beings moved into this area and began cultivating and storing great quantities of seed crops, house mice seized the opportunities offered by this beneficent species. They built their nests in the predator-free environments they found in barns, sheds, and houses and learned to eat seeds that had been processed into bread, as well as whatever other foods human beings left lying around. A few house mice traveled with human beings wherever they went and established new populations to take advantage of the opportunities offered by new settlements.

House mice have been able to keep pace with their human companions because they are physically adaptive enough to survive in most of the habitats human beings can survive in, and they are behaviorally flexible enough to mate and reproduce wherever they find themselves. The only places house mice don't do well are where they must compete with established populations of wild native rodents and where they are confined with too little food to support their numbers. Given an absence of competition from other rodents, ample food, and opportunities to disperse as necessary, house mice set up social systems designed to populate and colonize whatever space is available to them.

In close quarters, adult males establish territories, driving out other males but inviting up to ten females to stay and mate. Each male patrols his territory regularly and marks its boundaries with his strong-smelling urine. The females, who become ready to mate in response to the scent of male urine, bear their young about three weeks after mating, each female producing four to seven offspring, who need care for about three more weeks. During this period, several females might put their young in a communal nest and take turns feeding them. When the young mature, many of them, especially the males, are driven out of the parental territory and become potential colonizers of the surrounding area.

If they are forced outdoors, these colonizers live according to somewhat different social patterns. The males are still aggressive toward one another, but they don't establish and defend their territories. Rather, the breeding groups move around as food sources change and new habitats open up to them. These different ways of living and breeding, settling and wandering, assure that house mice will move into whatever spaces become available to them.

Because house mice prefer exactly the same foods we do—with the exception that they also eat insects—and because they often try to settle in our homes, they are direct competitors. Therefore, if a family group tries to establish itself in your kitchen range, you must respond to protect your interests. If I had known a little more about house mouse habits, I would have done a little more than replace the lining and set half a dozen snap traps. For instance, I would also have put all my food in tin cans or glass jars, rearranged my kitchen to confuse and disconcert the resident mice, and blocked the holes they were using to get in and out of the range. Changing the environment in a kitchen is a much more effective way to eliminate house mice than leaving the environment tempting to this opportunistic species and trying to trap all the new individuals who will find their way in. [13]

THE PORCUPINE PROVES ITS POINT

Though people and porcupines have lived together for centuries, the ubiquitous rodents remain mired in misconception. Some researchers now believe, for instance, that the porcupine's ability to destroy trees is overrated. What's more, its reputation as the stupid slug of the mammal world clearly is wrong, and our understanding of how the animal defends itself often is incorrect. And rather than being an impregnable fortress, the porcupine is susceptible to predation by a small carnivore that few people ever see.

The porcupine waddled north from the South American tropics at least 2.5 million years ago and now lives throughout most Canadian and U. S. forests (with the notable exception of the southeastern and prairie states). In addition to North America's one species, several more of the same family exist in Central America and South America. The name "porcupine" comes from the Latin words for "swine" and "thorn." Some people simply call them quill pigs.

The continent's second-largest native rodent (next to the beaver), an adult porcupine averages 30 inches in length and weighs 15 pounds. It has a small head with blunt snout, prominent nostrils, and large incisors well suited for gnawing. A muscular, 11-inch tail provides balance for climbing trees.

It is the porcupine's 30,000 quills that make it unique, however. Each quill is a specially adapted hair made stiff by its cylindrical shape and the spongy material at its core. In quiet moments, these hairs lie peacefully against the animal's skin. But when a porcupine becomes frightened, it raises the quills to create a formidable and potentially lethal pincushion of defense. Quill length ranges from less than an inch to more than four inches, and only the animal's nose, throat, and belly are not protected. Contrary to popular belief, a porcupine cannot throw its quills, but it does flail its tail, frequently driving quills deep into the aggressor.

Once stuck in a victim's skin, the quills detach easily from the porcupine and are replaced in a few weeks. Each quill's needle-sharp tip possesses a multitude of microscopic, backward-pointing barbs, which make extraction painful. If a quill is not removed immediately, its spongy core absorbs body fluid, expands, and causes the barbs to flare outward like tiny arrowheads. Each time the victim's muscle contracts, the quill penetrates farther and may travel through the body. Many predators—even a few humans—have died when quills punctured vital organs.

Porcupines have evolved one of the slowest reproductive rates of all mammals. The female becomes sexually mature during her third year and bears only one young annually for the remainder of her fifteen-year life span. By comparison, another rodent, the North American meadow vole, breeds every twenty-five to thirty days and may produce more than one hundred offspring each year. Because the porcupine's excellent defense results in low mortality, it doesn't need to reproduce rapidly to keep its numbers up.

[14]

WHERE HAVE ALL THE WOODCHUCKS GONE?

When people think of woodchucks, their thoughts are probably not kindly ones. Certainly, someone who has had his or her garden raided by a woodchuck has less than fond thoughts of an animal which has just eaten all the snow peas. Probably few people are thinking about woodchucks at this time of year. They are not to be seen because they have retired to their underground home, where they are hibernating.

The time of entering and emerging hibernation is affected by temperature and general latitude. On my farm, woodchucks often begin hibernation in late September and awaken in March. Some woodchucks may emerge from hibernation during warm weather in midwinter, but when it gets colder, they resume their hibernation. It is a myth that if a woodchuck sees its shadow on February 2 there will be six more weeks of winter. This myth was most likely originated by European settlers. The European hedgehog (not a close relative of our woodchuck) was thought to be a predictor of the weather.

A woodchuck does not store food in its underground burrow. Instead, it eats an enormous amount of food during late summer. It lives off stored energy, which is in the form of three-quarters of an inch of fat. The woodchuck's energy demand is low during hibernation. Its heart rate drops from one hundred times per minute to four times per minute. Its temperature drops from above 90°F to the low 40s. The woodchuck reduces its breathing to once every five minutes.

Most of the fat reserve actually gets used up when the woodchuck emerges in early spring, when food is scarce. The males emerge first. They wander around in search of a female. Females are choosy and will drive a male away.

A month after mating the female gives birth to two to six young. The baby woodchucks are born blind, hairless, and helpless. They are nursed for approximately four weeks. Weaning begins with the female bringing them green food. At five weeks of age the young begin to feed on their own, and by August the mother pushes them away to build their own burrows.

Woodchucks choose sandy soil on a well-drained hillside in which to build their burrows. This helps prevent flooding of the den. The front feet, with four well-developed claws, are used to loosen the soil. The woodchuck uses its powerful hind feet to kick the soil backward. Like all rodents, the woodchuck has powerful incisors, and these come into use in cropping roots.

A single woodchuck can dump seven hundred pounds of earth in one season. This helps improve the soil, since soil which is brought to the surface gets beneficial weathering. Abandoned woodchuck burrows provide housing for other animals. This may not help endear woodchucks to us. Perhaps the thought that they are food for a multitude of animals, including foxes, bobcats, large hawks, and coyotes, might be more appreciated.   [15]

SKUNKS

The skunk's restless search for a mate
Affects all of us, sad to relate.
His love signals are scented;
One knows where he wented.
Best retreat before it's too late.

Your nose can tell you spring is coming; in fact, the message may already have gotten to you, strong and unmistakable. Skunks are out! During warm spells throughout the winter, skunks, especially the males, prowl in search of food, but at this time of year the search is for mates.

The mating season lasts from mid-February through March, with the older females coming into heat before the younger ones. Four to ten young are born in nine weeks, blind, toothless, and mouse-sized. In seven weeks they are weaned, but their hunting lessons begin before then, as they follow their mother about looking for beetles, grasshoppers, and other insects. A Department of Agriculture study showed that more than half the bulk of 1700 skunk stomachs examined consisted of insects. The rest was composed of mice, earthworms, turtle eggs, fruit, and grain.

Skunks are nocturnal. They can't see very well, but their sense of smell is excellent. One would think an all-black coat would be the best camouflage color for the night, but the skunk has no need to hide the white marking on its back (usually a Y-shaped pattern with the top of the Y framing the tail). In fact, that warning coloration protects it from enemies who have learned the potent defense of the little black nighttime wanderer. A skunk's spray can be quite devastating, temporarily blinding and nauseating its attacker, but it usually gives fair warning with stamping front feet and raised, fanned-out tail. Then watch out! Still facing the attacker, its rear end whips around to let loose. Accurate up to ten feet, the skunk can shoot from one or both of the ducts underneath its tail and repeat the discharge up to six times if necessary. It usually isn't.

Automobiles and great horned owls, neither of which has a very good sense of smell, are the skunk's two main enemies.

I'm not crazy about the smell of skunk, but I am grateful for its reassurances that spring is on the way.　　　[16]

RED AND GRAY FOXES

Appearance is not the only distinction between red and gray foxes; their preferred habitats are quite different too. Gray foxes usually choose to live in rough and rocky terrain in dense hardwood or mixed forests, swamps, and thickets. Red foxes, on the other hand, prefer a mix of open and forested areas within their habitats. They thrive in the rolling farmlands and broken mixed hardwood forests. It is intriguing to me that gray foxes can climb trees but not red foxes.

The denning requirements of the red and gray foxes are distinctive. While gray foxes almost always make their dens above ground—in cavities, hollow logs, rock crevices, or beneath deserted buildings—red foxes choose to den below ground. Red foxes may dig their own dens, but they usually prefer to enlarge the former dens of woodchucks or other mammals. Red fox dens are complex tunnel systems, twenty to seventy-five feet long, constructed at least four feet below the earth's surface. The dens often have several entrances— usually one or two main entrances and two or three less conspicuous plunge holes—designed to evade potential predators. Red foxes, as well as gray, often maintain at least two separate dens so that, if one den is disturbed, the young may be moved to the other.

The ways that red and gray foxes raise their young are quite similar. The breeding season for red foxes runs from mid-January until late February, peaking in late January; while gray foxes breed from mid-January until May. Prior to the birth, the male is chased from the den area, and the female prepares a soft, grass-lined maternity chamber within the den. Red fox vixens give birth to an average of four or five pups per litter, while gray foxes usually bear only three or four young. During the first weeks of life, the vixen tends to the young alone, remaining in the safety of the den and nursing the pups with milk three times richer than cow's milk. At two weeks for the red fox or three for the gray fox, the vixen welcomes the male back to the den, at which time he begins to share in the responsibility of gathering food, which is first regurgitated and later simply given to the pups at the den site.   [17]

WHITE-TAILED DEER: ADAPTING TO ALL SEASONS

The ancestors of the white-tailed deer (*Odocoileus virginianus*) were not native to America but emigrated from Asia by crossing the land bridge millions of years ago. The white-tailed deer evolved as a species in North America and, because of its ability to adapt to changing conditions, continues to thrive.

As a prey animal, the deer is well suited to detect and avoid its predators. A phenomenal sense of smell, plus acute hearing, warns a deer of trouble often before the predator, man or beast, knows of the deer's presence. Its eyesight is also very good, quick to notice the slightest movement. Once a predator has been detected, the deer may choose to run, hide, or fight. It is well equipped to do any of these. In short bursts a deer may run faster than 30 m.p.h., and leap more than 25 feet over obstacles as high as 8 feet. Its uncanny ability to hide comes partly from an intimate knowledge of every nook and cranny in the approximate square mile it inhabits, plus the camouflage of a bark-colored, brown-gray coat for the nonsummer months. As a fighter, a deer can attack with very sharp front hooves.

The deer's digestive system also contributes to its ability to avoid predators. As a ruminant, with a four-chambered stomach, it can eat and run, storing the food in its first stomach. Later, when there is quiet time, the deer will regurgitate it, chew it as cud, and pass it on through the last three stomachs for final digestion.

The life cycle of the deer is closely related to seasonal changes. Fawns are born in late spring after a two-hundred-day gestation period. A healthy doe will give birth to one fawn her first year and usually twins from then on if she has adequate food. Within minutes the fawns can walk, suckle, hear and react to noises, and lie absolutely still when commanded to do so by their mothers. With their spotted coats and absence of odor for the first few weeks, they are almost undetectable unless they move. Fawns grow rapidly on the rich doe milk, which has almost twice the solids of cow's milk and nearly three times the protein and fat. By fall, at the age of about four months, they are weaned, and only their somewhat smaller size and absence of antlers distinguish them from adult does and bucks. [18]

LYME DISEASE AFFECTS BOTH HUMANS AND ANIMALS

Although ticks have never been a real problem in Vermont, they can be found in some areas of the state. In addition to being a nuisance, many species may carry Lyme disease. This disease, named after a 1975 epidemic of skin irritation and joint pain in the town of Old Lyme, Connecticut, affects both animals and humans. Common signs include fever, lack of energy, swollen joints, and arthritis.

The deer tick can transmit the disease in any of its developing stages, although several other ticks, including the American dog tick, also may be carriers. Although once commonly misdiagnosed, increased awareness and improved diagnostic methods in both humans and veterinary medicine have helped diagnosticians accurately diagnose Lyme disease.

Dogs most often are infected, although the disease is also commonly seen in humans and horses. Canine symptoms include a sudden onset of lameness with pain, swelling, and heat in one or more joints. The lameness usually lasts only a few days, but about one-third of affected dogs experience chronic lameness. The foreleg is most commonly affected, but in two-thirds of dogs with Lyme disease, two or more joints are involved. Skin lesions, often seen in humans, have not been reported in dogs or other animals.

Lyme disease is treatable with antibiotics. Early detection and preventive treatment decreases the chance of the dog's developing cardiac, neurological, or arthritic complications. Treatment during the later stages of the illness requires more-intensive therapy.

Since a vaccine is not yet available, the best way to prevent this disease is to check pets periodically (especially after walking in tick-infested areas) and to remove any ticks found. Ticks must be removed intact, as leaving the mouthparts in the skin can cause irritation and infection. Spraying and dipping pets can remove undetected ticks. Tick repellents also are available for pets—and their owners.   [19]

LIMITS TO GROWTH

Predation is one of several factors which affect prey populations. One important concept to keep in mind when examining population changes is that of carrying capacity. The carrying capacity for a particular species in a particular place is the number of individuals that the resources of the habitat can support in a healthy condition over a relatively long period of time, *without damage to the environment*. It can be thought of as an ultimate environmental limit.

Given a maximum limit such as this, why would some populations of birds and mammals fluctuate so widely and regularly? The lynx population in the arctic Hudson Bay region of Canada, for instance, closely follows the roughly ten-year cyclic population changes of the snowshoe hare, its principal prey.

Since lynx feed primarily on hare and the hares have few other predators, the cause of these cycles would seem to be the regulatory effects of predation. An increase in the hare population makes more food available for the lynx, so the number of lynx increases. This greater number of lynx exerts a higher predation pressure on the hare population, which eventually is high enough to outstrip the hare's ability to reproduce. Hares become more scarce, and lynx have difficulty finding those that remain. In time, lynx become malnourished and die of starvation and disease. This relieves the pressure on the hare population; more young and adult hares survive to reproduce, and the cycle begins again.

This is a simple, sensible explanation—but as it turns out, this model is not adequate. According to Dr. Robert Ricklefs in *The Economy of Nature*, "the reproductive potential of the hare is so much greater than that of the lynx that the lynx population could not increase fast enough to exterminate the hares unless some other factor, perhaps insufficiency of food, slowed the growth rate of the hare population." Besides this, lynx population peaks occasionally coincide with or precede hare population peaks, rather than following them by a year or so; and on some islands where lynx are absent, hare populations fluctuate just as much as they do on the mainland.

What might be going on here? The carrying capacity of the hare's habitat, rather than predation by lynx, may be the factor controlling its population size. The increased numbers of hares deplete their own food supply and thus assure their own decline.   [20]

COYOTE

Around the turn of the century, when the cattle and sheep farmers of the western plains turned to the federal government for help in the slaughter of wolves and coyotes, eighteen thousand coyotes were killed under federal bounty in a single year. Two coyotes, though, found a way to survive, at least for a while. There was at that time, roaming about in the Wyoming–South Dakota area, a legendary wolf of such cunning and daring that it defied the best efforts of bounty hunters sent to kill it. It was known as the Custer wolf. What made the Custer wolf so difficult to kill were two sentinels that traveled on each side of it—two coyotes that accompanied the wolf everywhere and shared in the wolf's great hunting success. Eventually the government mobilized. The Custer wolf was tracked relentlessly, its coyote companions were shot, and then six months later the Custer wolf itself was killed.

The story is instructive. Wolves and coyotes are not allies—indeed, wolves have been known to kill coyotes—but if ever an animal has been able to adapt to adverse conditions, it is the coyote.

Coyotes can run at thirty miles an hour, sometimes faster; they can leap up to fourteen feet; coyotes are strong swimmers; and they are excellent hunters with superior vision, smell, and hearing. But perhaps their greatest asset is their varied diet. They are opportunistic feeders, able to eat anything from grasshoppers to carrion. Other large North American predators, such as the wolf and the mountain lion, were unable to drop down the food chain and eat foods other than their preferred prey. Because the coyote is able to make do eating vegetation, it has been able to survive and prosper in an ecosystem that destroyed other predators. Coyotes have another advantage. When pressed, the species adjusts its breeding capability. Normal coyote litters consist of three to four pups, born in late April or May. But when the coyote population in an area diminishes, the female produces larger litters of seven or more pups.

In his book *In Search of the New England Coyote*, Peter Anderson tells of meeting Ben Day, Vermont's director of wildlife, while in search of coyote lore. Day told him many coyote stories and concluded with this thought: "I don't know who will inherit the earth. If not the meek, maybe it will be . . . the coyote."   [21]

A WOLF SOMEWHERE SINGS

A furry, dark-rimmed ear twitches. Again. Two yellow eyes emerge to peer from beneath recalcitrant lids. Snow. Fairly large flakes of it. The head behind the eyes slowly lifts itself from two large canine paws rested parallel upon the earthbed. A nice nap, but it's over. Something in the air besides snow.

She yawns. All forty-two well-adapted, classically carnivorous teeth show themselves to the winter world, a world that, should nature concede, will again provide her with warm muscle meat and fatty tissue on which to use those teeth. To endure life. To procreate. To recreate.

It is a good, worthy yawn, complete with the audible whine for closure. Instantly her ears perk up, eyes widen. A face capable of a thousand definitive expressions shows one of caution, caution and embarrassment, as if to say, "A foolish thing, that noise; what will come of it?"

The wolf stands. A bit larger than average for a female of her species: thirty inches at the shoulders, five and a half feet between tips of tail and nose. All seventy pounds of her two-and-a-half-year-old form celebrate the occasion in a ceremonious stretch, bowing to the earth.

Beneath her fur she is warm and dry. It is a beautiful and marvelously adapted coat, now two and a half inches deep. Thickest now because of winter, it would defy a finger's complete penetration to the skin below.

Another yawn, this one nearly silent. A thin layer of

snow has accumulated upon her back and head. Whether from this or some other motivation for which she as wolf cares little to contemplate, a quivering shake, emanating from her head, virtually rolls through her thick neck and down along her body. Snow flails in every direction.

Move. Into the woods, straight across the meadow and through the stream, pausing midway for a good lap of cold water, then on into more woods. Five miles per hour is her trotting speed, one which can be easily maintained for hours. Movement is for her, of course, an occupation—her liveli-hood. She also seems to enjoy it. She'll cover up to thirty miles in a day, in snow. Some of her kind have been known to travel 125 miles between midnights. Her relatives to the north will at times follow caribou five or six miles before initiating an attack, then, at speeds nearing forty miles per hour, pursue their target. Speeds over twenty-five miles per hour can be sustained, if necessary, for upwards of twenty minutes.

Reaching a snowdrift, she leaps over the crest and bounds through the three-foot depths for some thirty feet before reaching the shallows beneath a conifer canopy. She then resumes her trot.   [22]

BLACK BEARS

Black bears are solitary animals that establish individual home ranges. A male's home territory may include 50–100 square miles and the female's 10–50 square miles.

In their natural, wild setting bears are extremely wary animals. They prefer seclusion, shun human presence, and will avoid human intrusion such as development and roads. From impenetrable spruce-fir growth on mountain ridges to the heavy vegetation along streams and wetlands and thick undergrowth in timbered woods, cover provides bears with critical privacy and protection. In areas with good cover, a bear's need for remoteness is reduced, but where cover is not optimal, a greater degree of remoteness is required.

Seasonal food is the most important factor to the bear's reproductive success and survival rate. The routes that bears travel throughout the year—"bear corridors"—are the critical link to food sources, and are also key during breeding season and when young bears disperse to find their own home ranges. Bears' food and travel needs are best understood by looking at the bear through the seasons.

About two weeks after leaving the den, the female's biological processes have returned to normal. Although she is still living off of her fat stores and will continue to do so into June, she will begin eating as well.

Wetlands provide some of the earliest spring food, when little else of nutritional value is available. Roots, tender shoots of grasses and sedges, bulbs of various herbs, and tree buds provide a diverse and abundant diet. Bears, though classified as carnivores, are true omnivores and will supplement this diet with leftover nuts, evergreen needles, and carrion.

A mother with cubs may have to travel long distances if spring food is scarce within her home range. This depletes her fat stores and may stress her body so much that she can no longer nurse; in turn, this can lead to the demise of the cubs. This is also true for sub-adult bears. The first year or two away from the mother are difficult, and extensive food searches increase the likelihood of mortality.

This critical time extends to mid-July, when many plants and "soft mast"—fruits and berries—become abundant. Shadberries, strawberries, raspberries, blueberries, blackberries, as well as various colonial insects, will be staples of a bear's diet throughout the summer.

June through July is breeding season. If they are of sufficient size and health, bears of both sexes mature at three to five years old. Optimally, females breed every two years (that is, a female with first-year cubs will not breed, waiting until the next year, when her cubs have left her). Males will travel out of their usual home ranges and may breed with more than one female.

In August, a female bear that has cubs will stop nursing,

POLAR BEARS

One of the largest carnivores in the world—adult males usually weigh more than 1000 pounds, females about 500 pounds—the polar bear is a prodigy of power and agility. Its massive forepaws, somewhat larger than its hindpaws, can measure 12 inches in diameter. These large front paws, with their partially webbed toes and long, wicked claws, enable the bear to swim rapidly, dig easily, and stun or kill a 500-pound bearded seal with one ponderous blow.

This Arctic heavyweight is an agile and crafty hunter. It can quietly lower itself backwards into the sea, swim underwater, or on the surface with only its black nose showing, and then burst from the water to attack its prey. It can run with surprising speed across rough ice and leap over six-foot obstacles. When stalking a seal resting on ice, the polar bear employs cat-like stealth: crouching low to the ice, sliding forward almost imperceptibly, stopping when the seal looks up from its nap, and finally rising to its hind legs for a swift and lethal attack.

Hunting bears are attracted to open-water channels, or "leads," at the edge of the pack ice, because seals and belugas often gather in these, where they do not have to break ice to come up for air. In the absence of leads, or seals on ice, the polar bear will seek out breathing holes maintained by its prey. When an unsuspecting seal or whale surfaces for air, the bear kills it with a crushing bite or a powerful blow.

The bears mate in April and May, on the pack ice. Pregnant females build maternity dens in snowdrifts in October and November. These are usually oval rooms about five feet high, five feet wide, and up to twenty feet long, with an entrance tunnel six feet long. In December and January, after an eight-month gestation, the pups are born—blind, helpless, and about the size of a guinea pig. The mother and her nursing cubs remain snug in their den for the next three or four months, emerging in March or April to begin hunting baby seals. The she-bear and her offspring will hunt together until the young bears are almost two years old. At that point, the mother usually abandons them.

Adult males take no part in rearing the cubs, and in fact pose a danger to them. It is not unusual for a hungry, irritable male to kill a cub and eat it.

The estimated world population is no more than twenty thousand, and possibly less. Of this number, the largest concentration—about twelve thousand—occurs in the Canadian Arctic. [24]

CATTLE COUNTRY

In 1624 three heifers and a bull, the first cattle in New England, landed at Plymouth. Six years later, in 1630, Governor Winthrop's fleet arrived with 170 head, the survivors of 240 loaded in England. Many of these cattle died in the bitter winter that followed, as did many Puritans, but in the spring of 1631 ships brought more people and more cattle.

Had the Puritans faced a dark and howling wilderness, located just beyond the high-tide mark, shiploads of cattle would have made little sense. In fact, the English found large fields cleared by the Indians for corn and thousands of acres of saltwater and freshwater marshland suited for cattle-grazing.

The English were a great cattle-raising people, and so quickly did the Puritans expand their herds that by 1650 over seventy towns, from western Connecticut to Casco Bay, had been established at locations with extensive marshlands; the Bay Colony itself, led by the cowtown of Cambridge, counted twelve thousand head.

The colonists raised other livestock as well, including a large number of free-ranging hogs and goats and small horses. But none approached cattle in importance. Cattle were important because, as ruminants, they were able to convert low-grade roughage to milk, meat, manure, hide, tallow, and draft power. Leather, a material we may take for granted nowadays, was then a vital ingredient of civilization.

But the prime job of the frontier cow was to drop bull calves, which, after castration, would grow up to become oxen. Oxen were stronger, tougher, and safer than horses, much cheaper to keep and outfit, and tasted better after retirement. Essential to the settlement of New England, they were the chief log-twitchers, stump-pullers, stone-draggers, sled-haulers, and plow-drafters for two hundred years. The thousands of miles of stone walls in New England are, in large part, their work, and the great colonial mast trade, centered in southern New Hampshire and southern Maine and active for over one hundred years, was run on oxen power.

Ancient mast pines, some standing well over two hundred feet, were considered of strategic military importance and were reserved by law for Royal Navy use. A typical mast log might be four feet through and forty yards long. Removing it from a swamp might require forty yoke of oxen; a bigger log might take twice that number.   [25]

SUGARELLE

She lived among others of her ilk on a river farm, just one of many, in a mixed herd of dairy cows. Mostly she tried her best to be a cow. She would suffer herself to be milked; she would walk with her sisters in a line to pasture, there to spend the day, grazing, chewing her cud, and staring into space in the manner of cows. And in late afternoon, she would take her place by the fence and, when the gate was opened, proceed to her stanchion and wait her turn to be milked. All in all, a normal cow.

But that was only the superficial Sugarelle. She was in fact half human. For some unknown reason, early in her calfhood, she had developed a strong affection for human company. It was more than an affection, it was an affinity, something deep within her soul that told her she was not cow but woman. Freed from her stanchion, if permitted, she would follow you into the yard; sometimes she would spend the day there, clipping the lawn, watching the cars come and go on the long dusty driveway, and following the packs of children that streamed to and fro between the barn and the main house. Periodically, on her pasture-free days, Sugarelle would attempt to come in with them. The door was too narrow to admit her, and she would stand, head inside the kitchen, shoulders butted up against the frame, chewing and staring and hoping for something.

Newcomers were "introduced" to her. She was allowed to graze on the lawn whenever the minister visited, and on holidays, when the family gathered for picnics.

As children, my cousins and I used to let her lick the salt from our hands and arms. We would scratch behind her ears, lean our cheeks on her warm, sunny flanks, climb up and lie along her back. But it was not for these curiosities that I remember Sugarelle. I remember her best as a normal cow.

To my mind her finest hours were when she took her place among her sisters. I loved to watch her file out to the distant pasture; I loved to see her stand in the shade of the great holly tree in the middle of the field, or see her waist deep in the farm pond, her noble head raised, eyes fixed on the distant horizon, lost in her cowly thoughts. At these times she seemed mostly what she was, quintessential cow, symbol of the ancient union between human life and animal life, the benign symbiosis, the cow mother, the creator, the source of life. [26]

MIRACLE AT MIDNIGHT

Thinking back to when we had our dairy herd, I recall that there aren't many noises in a cowbarn at midnight, but those that do break the silence are memorable. They are as much a part of a barn atmosphere as the smell of hay or the sight of cobwebs. There are few occasions when a farmer has a chance to hear his barn in the middle of the night. He will be sleeping when it is quietest—after the cows have their fill of hay and are dozing in their stalls.

Thus, the need to sit up with a cow when a difficult calving is anticipated can offer an unusual opportunity to experience the stable outside its rush hours. There are those of us who like the chore.

There will be the occasional clinking of a stanchion chain as a relaxing producer shifts her body weight, and her stomach may growl in fermenting pleasure as the paunch rolls clumsily. Her stomach may be the noisiest thing in the barn.

If we wait awhile, "Old Brindle" will throw her head backward to lick an itching rib, curling the hair with artistic sweeps of the tongue and appearing to savor the dandruff. At pasture, she would use a fence post or tree to handle the itching.

But the undertone of sound that runs through the herd is something the listener will miss if there are distracting movements within the herd. It is the rhythmic chewing of cuds as the evening's intake of hay is methodically ground.

Often, the waiting husbandman will notice the cud-chewing only when it stops and the cow produces her unmannerly swallowing sound that bovines are not taught to subdue.

Even the calving is a quiet event with softly toned laborings. Only when the newborn is stretched on the straw will the cow arise and utter her low, insistent calls of motherhood as a sandpaper tongue clears membrane from flaring young nostrils.

The birth is so quiet that the rest of the herd chews on without the turning of a single head.   [27]

TRAINING OXEN AND STEERS

Every steer should have a name. If he has none, be sure to give him one the first time he is yoked. Be sure and make each one understand his own name and know when you speak to him, and when you do speak, say just what you mean. Be just as particular in your language with them as you would be with children. When you tell them to "haw-to" or "gee-off," be sure to make them mind you. Let the word "whoa" or any other word you may choose to substitute denote "stop," and always make them stop at that and never use it at any other time. If used at other times, they will not know when to stop. A team should stop short at the word "whoa," and they will if they never hear it at any other time. Bad results may follow their not being accustomed to doing so.

If too lazy or too tired to walk beside your team, never whip them while riding; this will make them haul apart. When your team is moving just as you want to have them, be sure to keep your whip and tongue perfectly still. When I see a man doing this, I know he is more than a middling teamster. Oxen or steers, when going perfectly right, should never be meddled with, any more than boys should be muttered at when they are doing perfectly right.

Always have some particular word to start your team with. My starting word is "come." I always give them notice, and when up to the bow and ready, I speak the word "come," and if either ox does not attend to his business at that word, he is sure to feel the whip. I once knew a man, a good man and a good farmer, and I presume that he thought he knew how to drive a team as well as anybody, but he never handled his off ox. He drove the nigh ox and let the off one go as he pleased. When driving in hay and when near to the barnyard, he would begin to cry "whoa, whoa" about as fast as he could speak. I once had the curiosity to count how many times he said it after he arrived at the barnyard bars, and it amounted to one hundred and thirteen times. Still the oxen increased their speed until they got into the barn and were prevented from going farther. Doubtless they would have gone in as well if he had not said one word and stopped as well because they could not help it. How can an officer command men unless he has a particular word for a particular movement? And how, I wish to know, can we expect oxen to understand better than men? [28]

CAMELS

When Jane and I were in Egypt, we found our first ride on a camel an ordeal of physical discomfort, aches, and pains. And to top it off, a camel dealer offered me five camels in exchange for Jane. (Jane tells me it was ten camels!)

But stop and ponder for a moment the camel, one of the world's most astonishing creatures:

- It is famous for both endurance and exertion.
- A camel's body temperature ranges from a low of 93 degrees at night to a high of 105, allowing the animal to conserve precious bodily fluids during the hottest hours of the day.
- Although a man will die after losing 12 percent of his weight through dehydration, a camel can survive the loss of bodily fluids equivalent to 40 percent of its weight and still quickly recover once it reaches water.
- Camel's milk is so rich in Vitamin C and other nutrients that it formed the exclusive diet of the nomadic Reguibat tribe in the Sahara.
- Long ago the Bedouin women learned to wash their hair in camel urine because it killed lice, refreshed the scalp, stopped any itching, and gave the hair an attractive sheen.

The domestication of the camel some four thousand years ago made possible the human penetration of the vast desert areas of the Middle East. Without the camel, the Arab world would have had a vastly different history.

The one-humped Arabian camel is a marvel of adaptation. When the desert is in its winter bloom and the plants have a high water content, camels can go several months without drinking. So efficient is the camel as a beast of burden that on short hauls it can carry up to one thousand pounds, more than an elephant. Because of the camel, the Arab world made no attempt to develop a road system of its own until recently.

The camel is a masterpiece of nature, a creature perfectly evolved to thrive in one of the world's most hostile environments. A savage sun and scorching winds generate ovenlike temperatures that can kill a person within a few hours. Everything about the camel is exquisitely designed to allow it to flourish in this hellish land. Thick, protruding eyebrows act as "sunglasses," while long eyelashes shield the eyes from blowing desert sand. The ears and nostrils can both shut tight for the same reason. The broad feet of the camel spread its weight and keep it from sinking deeply into the sand. Most animals would starve on its diet of saltbush, spiky thorns, and sunbaked acacia leaves. A hungry camel will also eat the bones of its own kind, leather bridles, and its owner's tent.   [29]

BIRDS

BIRD LEGENDS AND LORE

Do you believe that eating mockingbird eggs will cure stuttering? Or that hanging a turkey vulture's skull around your neck will cure a headache? Or that eating the heart of a kingbird, raw, will cure heart disease?

Through time, and through different cultures, myth and folklore have been strong forces guiding mankind's beliefs and actions. Birds have always played a central role in superstitions, whether they are cast as gods, prophets, supernatural beings, luck charms (good or bad), messengers, weather forecasters, or powerful medicine.

Birds were used to explain away the otherwise inexplicable. Our ancestors were convinced that birds were associated with the spiritual world because of their unique ability to fly and to "disappear" into the sky. Early legends regarded birds as gods or, if not as actual deities, then as messengers to the gods. For ancient Egyptians, the falcon was the sun god because it was seen flying higher in the sky than any other bird. The vulture was thought a goddess who brought magical protection in the afterlife. The mummy of King Tutankhamen was adorned with the vulture goddess across its chest, and when Jane and I entered his tomb we found the falcon god depicted on various ornaments around the tomb.

The early American settlers brought with them many of the European legends. One such legend cast black-colored birds as evil. A crow that croaked three times as it flew over a house spelled death for one of the inhabitants. The raven predicted death or catastrophic events by calling out.

Slaves on southern cotton plantations thought the blue jay was the devil's messenger. The jay was a spy that understood and spoke human language. According to their legend, the jay would listen all week to the stories of man's activities and would report the stories directly to the devil himself on

Friday. The blue jay was happy and noisy during the week but never seen on Friday.

American Indians also endowed birds with powers of speech. They believed not only that the ruby-throated hummingbird spoke, but that it always told the truth. Parts of these legends live on today every time you hear someone say, "Oh, I already knew that; a little bird told me."   [1]

BUT FOR THE GRACE OF—FEATHERS?

Of all forms of life found on earth, birds are the only creatures that possess feathers. Researchers have found, after much laborious effort, that the lightness and insulating qualities of down, or body feathers, cannot be duplicated with other materials. How did these fascinating tools develop?

Scientists believe that reptiles and birds were once closely related. Through thousands of years of evolution, the characteristic reptilian scales evolved into a unique covering known as feathers. Like claws, scales, or fingernails, feathers are composed of keratin and serve several valuable functions.

The first is insulation. Birds live in virtually every habitat on this planet—deserts, oceans, temperate regions, the tropics, and polar regions. Their feather covering offers them the remarkable ability to regulate their body temperature to fit their particular environment. Many birds also have an oil gland at the base of their tail and regularly preen the water-repelling oily substance throughout their feathers to protect their bodies from the cooling effects of the water.

Feathers also play a part in camouflage. A bird's feather coloration generally mimics the habitat in which it resides. Many birds exhibit "countershading," or dark coloration on the back and light on the belly. This is also characteristic of many mammals and makes the animal more difficult for predator or prey to locate.

Feathers also function in courtship. For those of you who have been to the zoo and had the fortune of witnessing a peacock fanning his magnificent tail for some lucky hen, you have seen firsthand how elaborate feathers can be. In several species males have specialized "show feathers" for the sole purpose of courtship displays. These feathers are dropped when courtship is completed. Species with "show feathers" include egrets and the birds of paradise.

Probably the most evident use of feathers is flight. With the aid of highly specialized feathers and body design, birds are unquestionably the most efficient fliers in the world. From the hummingbirds who beat their wings seventy-five beats per second to the hawks who can soar effortlessly on hot air thermals for hours, the flight capabilities of birds are no less than miraculous.

Feathers themselves are highly complex and contain a number of intricate parts ranging from the vanes—the two flexible "sheets" on either side of the shaft, which is the part that connects it to the bird—to the barbs and barbules which make up the zipperlike system which interlocks to form an extremely strong and flexible unit. The parts of a feather may add up to more than a million individual units, many of which are so small they can be seen only with the aid of a microscope.

For anyone who has wondered how many feathers the average bird wears—yes, researchers have counted; and here are some of the numbers they have found:

Mourning Dove—2,635
Barred Owl—9,206
Red-bellied Woodpecker—3,665.

The greatest number was tallied on a tundra swan who sported a whopping 25,216 feathers! [2]

BIRDS AND THEIR REFLECTIONS

Nothing takes the belligerence out of a bird like window-cleaner spray. (Spray the chemical on the window—not on the bird!) The recipe has reference to those birds that devote much of the day to pecking on a cellar window. Or perhaps pecking on the hubcap of a parked automobile. In the hub-cap case, spray the gunk on the hubcap.

What's in motion in these cases is the bird's response in an amatory season to male intruders. Seeing himself reflected in the window or hubcap, the bird immediately attacks the image. And, as the bird can see quite clearly, the image is pecking back. In fact, the image keeps striking the bird's bill right on its point.

The window spray referred to is any type that lays a white layer on the glass or hubcap surface. It blots out the reflective quality of the glass. In addition, when it is rubbed off, the window spray gives the house the most beautifully clean cellar windows in the neighborhood.

The bird most likely to be pecking at himself in the cellar window is the male robin. The robin's involvement is quite simple. First, his turf is down there at cellar-window level. Second, the robin is a possessive, aggressive, and territorially exact bird. Third, robins view with favor the foundation plantings around the suburban home. The nest probably is within some evergreen planted close to the house.

If the bird attacking a window is not a robin, the next possibility is the mockingbird. On a scale of 100, 95 window-attackers are robins. Four are mockingbirds. The one to spare is a male cardinal. The reason the mockingbird and cardinal are among window-peckers is the foundation plantings. Both mockingbirds and cardinals use such shrubbery as nest sites. Perhaps the fact that they are not as prone as robins to be scratching a living out of the lawn accounts for their lower level of involvement with windows. They spend more time well above the earth's surface than robins do.

The window-pecking gambit is an activity quite apart from the once-in-a-lifetime crash into the picture window. That fatal maneuver too is caused by reflections, but not a reflection of the bird. Apparently wild birds see reflected in picture windows the woods and greenery that are at the lawn's edge. To the bird, the window appears an open area, since the lawn is the nearest reflection of the glass. The bird flies toward the trees in the background of the reflection and smashes into the glass. Such losses can be prevented by placing a net over the window. [3]

BIRD NESTS

John Terres tells of an old lady who shook out of her bedroom window every spring a fistful of feathers from her pillows. After their first surprise the birds seemed to welcome her offering, and every year they would fly up and look expectantly through the window. What will happen when poly-this and poly-that take over from feathers? Will birds learn to adjust to polys?

We heard of a pair of orioles that plucked hair from the head of their benefactor. This appears to be going too far, but you might put out lengths of bright ribbon or string for them. Not too long though, for the orioles might entangle themselves; six or eight inches will do. Hang yarn, thread, or string over a branch and see who takes it.

Horse hair is greatly prized by many perching birds. Persuade your horse-loving adolescent to save some mane- and tail-groomings for you. And hair of the short fluffy kind is a favorite for nest lining.

If you think you have some flycatchers about, hang out some strips of cellophane for them. They insist on weaving discarded snake skins into their nests as protection against predators but will use your substitute. We have a relative who each year brings up a hank of Spanish moss from the south and hangs it in her Cape Cod pine trees, a thoughtful gesture much appreciated by birds that are building nests.

There is also the true story of a greedy robin who was found to have incorporated a ten-dollar bill in her nest. You can encourage the less expensive tastes of robins, wood thrushes, and phoebes by having some mud available, perhaps in a shallow pan. They need mud for plaster.

We weren't patient enough nor possessed of nimble enough fingers to take apart the red-eyed vireo's nest that we had found our first fall. At a nature center someone did, separating all the materials the vireo had carefully fastened together and labeling and mounting them as an exhibit. The list was impressive: birch-bark bits, vine fibers, spider-egg case, plant down, bald-faced hornet's nest, bark fibers, bark/dirt, spruce twigs, wood chips, grapevine. It makes me tired just to think of collecting all that.

But the work of one European redstart must be the world's championship in collecting. Someone counted 361 stones, 15 nails, 146 pieces of bark, 14 bamboo splinters, 3 pieces of tin, 35 pieces of adhesive tape, 103 pieces of hard dirt, several rags and bones, 1 piece of glass, 4 pieces of inner tubes, and last, but not least, 30 pieces of horse manure! One wonders where the eggs were fitted in.   [4]

NESTS, NESTLINGS, AND CHICKS

Ancient birds probably laid their eggs in depressions on the ground. Nests constructed of branches or twigs apparently evolved after use of ground nests or natural cavities (holes in trees). Today bird nests are as varied as their occupants. Certain birds make scrapes on the ground or on cliff ledges. Nighthawks, sandpipers, short-eared owls, northern harriers, and grouse lay their eggs in simple depressions on the ground. Low vegetation or her own cryptic coloration helps to conceal the incubating female. Because predators have easy access to ground nesters, most lay numerous eggs to ensure survival of more of their young. Peregrine and prairie falcons make their scrapes on cliff ledges, which limits predation.

Canada geese usually nest on the ground, with females constructing simple nests of grasses surrounding a central depression to hold the eggs. Canada geese will also nest on structures, including old osprey nests and piles of hay. Trumpeter swans form their nest mounds on ponds, making access by ground predators difficult. Female mallards frequently nest at the edge of ponds, lining the nest with their own down feathers.

Some species prefer to nest in holes. Kingfishers and bank swallows dig burrows in river and stream banks and lay their eggs in these burrows. Woodpeckers use their chisel-like bills to drill nest cavities in rotting trees. Other cavity-nesters, such as red-breasted nuthatches, tree swallows, and bluebirds, occupy old woodpecker holes.

Red-tailed hawks and goshawks use twigs and branches to build big platform nests in trees. Osprey and bald eagles build very large stick nests, almost always adjacent to water. Osprey nests are usually built on the top of a snag (standing dead tree), but ospreys also make use of human structures such as power poles, duck blinds, and channel markers. Bald eagles generally construct their huge nests in the highest crotch of a large tree.

Ever see an owl building a nest? You won't, because owls do not make their own nests. Small owls such as saw-whet owls use woodpecker holes, while larger owls such as great horned owls appropriate an old hawk, crow, raven, or squirrel nest or find the right-sized natural cavity in a tree.

Female hummingbirds make tiny nest cups of seed down, mosses, or lichens, using cobwebs to bind their nests together. Songbirds, the most advanced of all birds, form many kinds of materials into a great variety of nests. Robins plaster their cup nests with mud and line them with soft grasses, and great crested flycatchers incorporate shed snakeskins and cellophane into theirs. Orioles hang their woven, pouch-like nests from branches, and Eastern wood-peewees camouflage theirs with pieces of lichens. Although tree swallows nest in old woodpecker holes, they avidly seek out feathers, preferably white ones, to line their nest cavities.   [5]

THE MOLT

Have you seen a crest-less blue jay or a crow with ragged wings recently? Both are showing evidence of a little-noticed aspect of birds' lives—the annual molt. Feathers are essential to a bird's survival: they provide mobility, insulation, camouflage, and identity. They are relatively tough structures but do become worn from normal use as well as from hosting ectoparasites and from being accidentally damaged.

Most adult birds undergo one complete molt each year. This usually takes place after the breeding season and before the fall migration, when food is still abundant enough to provide the energy necessary to grow a complete set of new feathers. Molt is usually symmetrical, to keep the bird's flying ability balanced; hence the ragged-winged crow has gaps in the same place in each wing. A few species, notably ducks, lose all their flight feathers at once and go through a period of three to four weeks unable to fly. The complete molt in most songbirds (wing, tail, and body feathers) probably takes five to eight weeks on average, the shorter time observed in long-distance migrants.

Several species of birds undergo a second molt, usually of just the body feathers, before they return for the breeding season. The scarlet tanager provides a good example of this. A male tanager in late summer has a pied appearance—that is, having patches of two or more colors—as red feathers are replaced by new green ones, which are themselves replaced by new red ones early the next spring. The black flight feathers are replaced only in the late-summer molt.

Occasionally birds are losing and growing so many body feathers at once that they have a decidedly ruffled appearance, which can be especially noticeable in birds visiting feeders and results in reports of crest-less blue jays and such in late summer.   [6]

THE SMARTEST FAMILY
OF BIRDS?

Are some birds smarter than others? No one knows for sure, but most experts agree that the blue jay and other members of the Corvidae, or crow, family seem to display the greatest levels of intelligence in the bird world.

"Corvids show greater ability to learn and are less restrained by instinct than other birds," says ornithologist Steve Sibley of Cornell University. "Their behavior is more imitative and plastic." In other words, they can adapt easily to a variety of environments and food choices.

Researchers have found that crows, ravens, magpies, nutcrackers, and other corvids have the largest cerebral hemispheres in relation to body size of all birds. They are known for their ability to manipulate objects, such as opening and closing latches and tying knots. In captivity, crows have learned to count and to read clocks, skills requiring keen eyesight and an ability to concentrate. In the wild, ravens use at least thirty distinct calls to communicate with each other. They can vocally imitate other animals, even human beings.

Corvids also appear to have remarkable memories. Blue jays can recall where they have cached thousands of acorns. It appears the birds cache near objects that to them are conspicuous, such as a tree in the middle of an open space or at the edge of some woods or hedgerows.

Though many of the bizarre stories related to corvid behavior are speculative, some are well documented. In Finland, for instance, ice fishermen have reported returning to their unattended tackle only to find hooded crows pulling in the lines with their bills. The birds back away from the ice hole and then carefully trot forward on top of the line to keep it from slipping. They repeat the sequence until they secure their catch. [7]

NASAL NOTES AND HEADFIRST HOPS

It's almost as though the bird were talking softly to itself, commenting about every seed and insect discovered as it skitters up and down the bark of the tree. "Yahn, Yahn, Yahn." This is the high-pitched, nasal call of the red-breasted nuthatch, which ranges from Canada south to northern Minnesota, Michigan, and northern New England through the Appalachians to North Carolina.

More familiar is the louder, deeper "Yank, Yank, Yank" call of its somewhat larger cousin, the white-breasted nuthatch, which ranges from southern Quebec and northern Minnesota south to Florida and the Gulf Coast. The bird's call does not sound so much like gentle conversation as busy-bodied gossip, pronouncing for all to hear.

The nuthatch family, *Sittidae*, probably got its name from Aristotle, who used the Greek word *sitte* to describe a bird that pecks at the bark of trees. The English word *nuthatch* comes from the bird's habit of hacking open seeds or nuts; hatch is a descendant of the word *hack*. Those two names, thus, describe major nuthatch characteristics: the typical feeding patterns of investigating bark crevices for insects and seeds and the ability to pry out and open up seeds and nuts with its beak.

Looking at these two birds, it is easy to see they are related. But as for their voices, there are differences. The red-breasted nuthatch is four inches long and is distinguishable by the soft rusty tones of its breast and belly. A close look also reveals a black stripe through its eye, bordered on top by a white stripe. Its back is gray-blue, its head black. The white-breasted nuthatch is larger—approximately five inches long—and has a white breast and belly. The whiteness of his underside, contrasting with a black and slate-gray back, gives it a crisp, clean look.

While both nuthatches are willing to share sunflower seeds at a feeder, they typically eat slightly different foods. The red-breasted nuthatch prefers seeds of pine, spruces, firs, and other conifers and is particularly adept at prying apart cone scales with its thin, sharp, pointed beak. The white-breasted nuthatch can handle larger seeds and nuts. It eats beechnuts and acorns along with corn and sunflower seeds. Both birds eat insects when available, especially during the warm months.

To watch these little birds zigzagging down a tree trunk headfirst in fits and starts—and to hear their nasal commentary—always cheers us up.   [8]

THE RIGHT TOOLS
FOR THE JOB

I enjoy watching woodpeckers in my trees. A hungry woodpecker doesn't have to grab a box of tools when it wants to dig into a tree trunk for food. It comes equipped with body parts perfect for the job.

While hammering, a woodpecker keeps its balance by bracing its stiff, springy tail against the tree. Without that stiff tail, it would be wobbly—as wobbly as you'd be if you tried to swing a baseball bat while standing on one foot.

A woodpecker's feet are built for clinging to the side of a tree. Most birds have three toes pointing forward and one backward. But most woodpeckers have two toes in front and two in back. The two back toes give the birds extra clinging power.

A woodpecker's beak is shaped like a chisel, a tool that people use for cutting or shaping wood or stone. With its beak, the woodpecker chisels a hole in the bark of a tree. Then it searches for insects with its tongue.

A woodpecker also has some mighty muscles that give "bang" to its beak. And it has extra-thick bone on the front of its head where its beak meets its skull. This bone is built like a sponge, with lots of air spaces inside. It acts a lot like the padding inside a football helmet. That's why the woodpecker doesn't get a headache from banging away on hard wood.

It's fun to watch a hungry pileated woodpecker at work on a tree. First, it listens for an insect grub chewing a tunnel through the wood. Then it starts chopping a hole near the sound. It aims first in one direction, then the other, just as a person does with a hatchet.

After a few seconds, the hole meets the grub's tunnel. Then the bird stretches its tongue wa-a-ay out and snakes it along the tunnel. When the woodpecker feels the tip of its tongue pass over the grub, it pulls its tongue back. The grub gets caught on the tongue's sticky tip, and the woodpecker drags the squirming insect into its mouth. [9]

THE VANISHING BARN OWLS

A pair of nesting barn owls and a brood of six young will eat approximately one thousand small mammals—mostly rodents—during the breeding season.

Given that aptitude for rodent removal, it's no wonder that the barn owl is a welcome guest at farmsteads throughout the world. In some countries, in fact, farmers routinely build special nest boxes under the eaves of their barns in an effort to attract the birds and thus reduce numbers of rats and mice on their lands.

Unfortunately, in America's heartland, where thousands of farms are found, all is not well these days for the "cats with wings," as barn owls are sometimes called. Though they continue to thrive in many areas of the United States, the birds are classified as endangered on the lists of seven midwestern states—Illinois, Indiana, Iowa, Michigan, Missouri, Ohio, and Wisconsin—and researchers are only now beginning to understand why.

As it turns out, the problem—and the best hope for a solution—can be traced to long-term agricultural trends that have resulted in the loss of grasslands and meadows, prize hunting habitat for the owls. That fact began to crystallize in 1980, when Bowling Green University's Bruce Colvin and researcher Paul Hegdal of the U. S. Fish and Wildlife Service launched a long-term study of thriving barn owl populations in southwest New Jersey.

Grassland is the key to the well-being of barn owls because it is prime habitat for voles, the widely distributed little mammals that are one of the bird's favorite foods. Colvin estimates that voles often make up about 70 percent of the barn owl's diet, even when other species of rodents are abundant.

To catch their prey, the owls have exceptionally acute night vision that enables them to pinpoint the rodents in near-total darkness. What's more, the birds' concave facial discs collect sound, such as the faint rustling of voles beneath mats of meadow grass, in the same way backyard satellite dishes pull in television signals. Uniquely structured wing feathers enable barn owls to swoop down noiselessly on unsuspecting victims. And at the decisive moment the birds use their long dangly legs to pluck prey from thick grass.

[10]

RAPTORS: ANOTHER COG IN NATURE'S WHEEL

Raptors are a highly specialized group of predators. They are characterized by large, strong toes ending in talons, used to grasp and hold prey. Birds of prey can be divided into five groups: hawks, eagles, falcons, owls, and vultures.

Our contact with raptors may range from the common sight of an American kestrel waiting patiently on a power line for its next meal to the rarer but much more impressive spectacle of a falcon in a two-hundred-mile-per-hour stoop or dive in pursuit of its prey.

Predators have long been considered a nuisance in man's grand design for nature. To this end, predators of all shapes and sizes have historically been eliminated as quickly as possible when their menus included domestic livestock or game species. A fact that is frequently forgotten is that man himself is a predator who puts much pressure on a few species for relatively short periods of time, while raptors exert a continuous and more subtle pressure on many prey species, whether the prey item is a grasshopper, a field mouse, or an occasional pheasant.

It is ironic that with the increased level of sophistication of wildlife biology, many of the basic ecological principles established by early wildlife scientists hold as true today as they did years ago. In a research study that is still considered a classic examination of a wildlife phenomenon, John and Frank Craighead, Jr., published *Hawks, Owls and Wildlife*, based on predator research done in the early 1940s.

What the Craigheads found was that raptors have an important role, or niche, in the environment. Raptors serve as continuous regulators of prey numbers, with the versatility to shift from one prey species to another if necessary or profitable. Raptors represent nature's answer to birth control, keeping normally prolific animals in harmony with limited resources. Raptors have another important role, serving as barometers of environmental quality. Being at the top of the food chain, raptors are generally the first group of species to decline because of environmental contamination by pesticides or pollutants resulting from man's attempts to control or change nature. Eggshell thinning in raptor species due to DDT accumulation is one example of how our contamination of natural resources has affected wildlife species. [11]

TERNS

Terns, or "sea swallows," as they are nicknamed, are a living presence of a summer day along the coast. Their slight, silvery form and earnest cries bring to mind the sea and sand more vividly than any other creature with which we share the beach.

Terns nest in dense colonies, employing a safety-in-numbers breeding strategy common to many seabirds. They nest on the ground and are therefore exceptionally vulnerable to predators. Massachusetts tern colonies range in size from a dozen or so pairs to more than three thousand.

Terns arrive in Massachusetts during the first two weeks of May and soon begin defending nest sites within the colony. After a brief courtship involving ritualized flight displays and intricate dances, mating ensues and is followed by the laying of two or three eggs, which are usually olive or beige in color, with dense brown splotches or speckles. The chicks hatch after twenty to twenty-three days of incubation and within a few days seek shelter in the beach grass or other concealing vegetation or debris. The chicks are protectively colored to blend in with their surroundings, as are the eggs.

Both adult and young terns eat only live food, mainly small fish caught in coastal shallows. Terns that are hovering and then plunging into schools of "minnows" are a common seaside sight in July and August. The preferred food is the slender, shiny sand eel or sand lance.

Terns can be aggressive when they are driving potential

predators away from the nesting colony, especially during the height of their breeding season, when they are defending their eggs and young. A beachcomber who unwittingly wanders too close to an active breeding colony is likely to be dive-bombed and "white-washed" by crying birds which are extremely agitated. The worst harm they can do is to give you a sharp rap on the scalp with their bills or wings, yet this may be quite uncomfortable for a stroller who is suddenly besieged by a mob of shrieking birds. If this happens to you, keep in mind that you have walked into the birds' nursery. Terns respond more or less as we would if our house and offspring were threatened. The best tactic is to make a hasty but careful retreat, taking care not to step on eggs or nestlings.

Parent birds continue to feed young for some time after they have begun catching their own food. By late summer large groups of terns begin to congregate on outer beaches and islands. Most depart for their southern wintering grounds in September.

Terns are among the world's great long-distance migrants. Some Arctic terns make a 25,000-mile, round-trip journey between the Arctic and the Antarctic via the coast of West Africa and South America. Most terns winter in the tropical zone of South America, but we still have only a sketchy knowledge of their wintering range and migratory routes. [12]

OUTRAGEOUS FORTUNE

Nay,
I'll have a starling shall be taught to speak
Nothing but "Mortimer," and give it to the King
To keep his anger still in motion,

Perhaps inspired by these words in Shakespeare's *King Henry IV*, Eugene Scheifflin introduced eighty European starlings into New York City's Central Park in 1890. Scheifflin, a bird watcher and influential figure in the American Acclimatization Society, thought that birds were scarce in America compared to England. His ambition was to introduce into this country every bird mentioned in Shakespeare. He tried many different species, but most didn't take. His two successes, however, were spectacular—the house sparrow, which he is thought to have established in 1860, and the starling, which is probably the most numerous bird in North America today. Shakespeare notwithstanding, people have been trying to get rid of both species ever since.

Today some urban roosts house 200,000 to 300,000 birds, and most newspaper stories about starlings deal with efforts to kill them or chase them away. In addition to competing with native birds, large congregations foul streets with their droppings, get sucked into airplane engines, injure fruit in orchards, and cause other, more unusual problems.

Playing tapes of starling distress calls drives them away for a while, but starling distress calls distress people, too. Standard approaches—shooting, trapping and gassing, installing electric wires or various noisemakers—have also been tried, but even when they work the surviving starlings just go bother someone else.

Scheifflin, it seems, succeeded beyond his fondest expectations. We may deplore the starlings' effect on native birds, but since we're stuck with them we may as well enjoy them. A starling in winter plumage—glossy black feathers iridescent with violet and green and sprinkled with beige specks—is a beautiful thing, and a flock wheeling in roller-coaster flight on a late-fall day is enough to make us understand why Scheifflin wanted so badly for us to see them. [13]

BALD EAGLES

In 1782, as thirteen American colonies were struggling to find an identity as the United States of America, Congress adopted the bald eagle as a symbol of the new country. Benjamin Franklin had suggested the turkey—a solid, utilitarian bird—but founding fathers such as Thomas Jefferson successfully backed the more dramatic eagle.

Unfortunately, from this exalted post as symbol of a nation, the bald eagle was forced into near extinction over the next hundred and seventy years. During the early 1900s, eagles, as well as other birds of prey, were felt to be nuisances, as they killed songbirds, small game, fish and, sometimes, farmers' livestock and poultry. They were hunted diligently, often with a bounty paid for each bird killed. Believing the eagle to be detrimental to commercial fishing, Alaskan bounty hunters alone destroyed more than one hundred thousand eagles in one twenty-year period.

Even with the passage of the federal Bald Eagle Protection Act in 1940, eagle populations continued to decline, but this time the killers were silent ones: DDT and related pesticides which were used in massive amounts during the 1950s and 1960s by the nation's agricultural sector. Residues of these pesticides accumulated in the nation's waters and fish populations, and, eventually, the eagles who relied heavily upon the affected fish as a food source began to die. In addition to dying from the accumulation of poison in their own systems, nesting eagles often produced eggs which were either infertile or so thin-shelled that they would break during the incubation period. Areas such as Chesapeake Bay, which, during the days of Native American habitation had been home for as many as a thousand nesting pairs, had as few as thirty to forty nesting pairs during this time of crisis.

Today, the bald eagle is felt to be on the road to recovery. A public uproar, stirred in part by the publication of Rachel Carson's classic, *Silent Spring*, led to the banning of DDT. The National Wildlife Federation took a leading role in developing public knowledge of the plight of the eagle and raising money to purchase eagle refuges. Given increasing levels of pollutants in the environment and continuing development of remaining wild areas, however, the eagle will probably always need special protection to enable it to survive. We hope that public awareness has been raised to the point where the bald eagle will remain a living symbol of freedom and strength. [14]

PEACE SYMBOL

Pigeons were among the first domesticated animals, tamed before chickens, turkeys, and other birds. Beginning in the Middle East more than four thousand years ago, people kept pigeons for food and soon discovered their usefulness as messengers.

The first record of pigeons as messengers dates from 1200 B.C., when four birds were released in opposite directions to carry the message of Pharaoh Ramses III's assumption of the throne. Nero used pigeons to send Olympic sports scores to his friends. In 44 B.C. Decimus Brutus sent messages via pigeon across the opposing lines of Mark Antony's army. Antony countered by putting up nets to capture enemy birds. During the Crusades, Christian armies intercepted Moslem pigeons by using a secret weapon, the peregrine falcon. By the twelfth century, though, the Moslems had established a regular pigeon post in Baghdad, perhaps the world's first airmail service.

Romans enjoyed pigeons as food, and their pigeon towers were probably forerunners of modern factory farming. These structures held as many as five thousand birds apiece, which were fed delicacies like bread softened by the chewing of slaves. A tower bird's life wasn't enviable, though. Keepers plucked its flight feathers or broke its legs to prevent escape.

In the Middle Ages, Romans brought pigeon-keeping to

other Europeans, who built ornamental houses lined with holes for nesting. From these we get the term "pigeonhole." By the nineteenth century, the British nobility had developed a passion for pigeon breeding and racing. Fanciers have since developed more than 150 breeds.

One breed, the Racing Homer, capable of speeds as fast as one hundred miles per hour and distances of six hundred miles a day, became invaluable for wartime communication. In both world wars, aircraft, tanks, and submarines carried homing pigeons. Soldiers suited some with tiny aluminum cameras for aerial surveillance photos. Other birds were airlifted across enemy lines strapped to the torsos of spies and released to fly back with coded messages.

In World War I, a pigeon called Cher Ami became a hero for saving an entire American battalion. It successfully carried a message across enemy lines, flying on even after one leg was shot off and bullets struck its eye and breastbone. After the war, Cher Ami sported a tiny wooden leg to replace the one lost in combat.   [15]

GOLDFINCHES

The male goldfinch is that small yellow bird with bold black markings. The usually unnoticed female is brownish. Goldfinches brighten their plumages each spring, just as most other birds do. Then they wait until summer's end seems in sight before taking courtship and family life seriously. The great ornithologist E. H. Forbush described the goldfinch lifestyle well:

> Panoplied in jet and gold, the merry, carefree goldfinches in cheery companies flit in summer sunshine. They wander happily about, singing, wooing, mating, eating, drinking, and bathing until July or August without family worries. . . . This vivacious little

finch is one of the most interesting and conspicuous birds of village, farm, and field. Its flashing yellow, its undulating, bounding flight, and its canary-like song have given it the name of wild canary among the country people.

What the goldfinch has been waiting for is the maturing of thistle seeds. The bird lines its nest with billowy thistle-down and gathers thistle seeds for food.

Goldfinches lack the urgency that the progression of summer seems to impress on many other birds. Many other birds must produce young early and strengthen them for a long migration to South America. But not the goldfinch. The goldfinch either is not going anywhere when frost arrives or is not going far. In winter goldfinches garb themselves in dingy feathers and search the same trees and fields that served them during summer.

In the days when ornithologists tried to justify the existence of birds on the basis of their value to humans, goldfinches ranked very high. Goldfinches are among the champions in consuming weed seeds. Their love of fluffy seeds has them searching for dandelion blossoms as well as thistle seed heads. Forbush notes:

> Two or three birds often may be seen hanging to the top-most branches of a slender weed until it bends to the ground under their weight, when they stand upon it and proceed to rob it of its fruition. It is so fond of the seeds of thistles that it is often called the thistlebird. Among other seeds of unculti-vated plants, it takes those of goldenrods, asters, wild sunflowers, wild clematis, mullein, evening primrose, dandelion, chicory, burdock, and catnip. Among cultivated plants, the goldfinch eats the seeds of zinnia, coreopsis, bachelor's buttons, cos-mos, millet, hemp, salsify, turnip, lettuce, and sun-flower.

That hemp to which Forbush refers is, of course, mari-juana. When he wrote some fifty years ago, hemp grew wild in the Missouri River valley, where the plant once was a prime crop. It was grown for rope fiber. Birds became quite frisky on hemp seed and at least mildly addicted. It was a simpler age, and few humans saw hemp as exhilarating.

[16]

HAWKS

Why is a group of hawks spiraling upward in the sky called a kettle? One can see these kettles in the fall over much of the southeast. Usually the birds forming the kettle are broad-winged hawks, a crow-sized cousin of the larger red-tailed hawks and red-shouldered hawks. Since broad-wings move on a rather broad front across New England, the phenomenon of the kettle of rising hawks may be seen almost anywhere. However, it is more likely to be seen above a flat, sun-heated valley that lies north of a steep hill.

The birds find a column of hot, rising air. They fly into it—often at a point less than fifty feet above ground surface—and rise effortlessly hundreds to thousands of feet in the air. While riding in a kettle, the birds keep their wings outspread and only rarely flap them. The air current furnishes all the lift. The occasional adjustment that the bird must make involves no more than a brief change of tail angle or a slight change of wing surface.

The bird, on reaching the top of the column of rising air, levels off and coasts away, heading southward. After a lift to the top of the kettle, the hawk rides higher than the hills immediately to the south. It soars over the hill range and coasts toward South America. After a few miles, depending upon how high the last column of air has lifted it, the bird must regain altitude. It accomplishes the task by looking for another air column and riding it. Thus broad-winged hawks expend little effort traveling from Quebec to Brazil. Instead, they ride an almost-endless series of roller coasters all the way.

[17]

THE BLACK-CAPPED CHICKADEE

Although the black-capped chickadee is a resident bird throughout the year in Vermont, it is at this time of the year that we really become aware of his presence, for now he may be seen about the house and, more particularly, at our feeding stations.

And it is also at this time of the year that we learn to appreciate this little bird, for when all nature seems to have retired before the cold blasts of the north wind, the chickadee is about, gay and happy, enlivening the winter scene with his amusing acrobatics and merry chatter of "chick-a-dee-dee-dee." He actually seems to enjoy a snowstorm, and even in the bitterest weather we find him frolicking from tree to tree, laughing and joking in his own inimitable way. He is the bird of whom Emerson wrote,

> This scrap of valor just for play
> Fronts the north wind in waistcoat gray
> As if to shame my weak behavior.

But in spite of his hardiness, the chickadee does require protection from cold winds and storms at night, finding shelter in some hollow tree or in a deserted bird's nest. And sometimes his search for food is interrupted for such a long time by severe storms, when the trees and shrubs become encased in ice, that even he must succumb to cold and hunger.

After the blustering winds of March have heralded the passing of winter, and even during the milder weather of winter, the chickadee begins to give voice to two or three high-pitched, sweet, and plaintive notes, the first being protracted and with a rising inflection, the others falling one to two tones lower, which sound like "phe-be" or "phe-be-be," which children of another day translated as "spring's come." And every now and then, some especially gifted male will attempt a jumble of slightly musical notes which seem to express the bird's unconquerable spirit and cheery disposition.   [18]

INSECTS

Hunt no living thing;
Ladybird nor butterfly,
Nor moth with dusty wing,
Nor cricket chirping cheerily,
Nor grasshopper so high of leaps,
Nor dancing gnat, nor beetle fat,
Nor harmless worms that creep.
　　　　　　　　—Christina Rossetti (1830–1894)

SECRET LIFE OF A FOREST

Watching for wildlife in the forest, we rarely see past the surface of things. Standing on the ground floor, we scan the leafy rafters—entirely overlooking the living world in the soil beneath our feet. The forest's basement is a secret world. As different from our own world as water is from air, the soil seems quiet, even dead. But life bustles down below: a cubic inch of topsoil may contain billions of creatures.

Predators and prey roam beneath as well as above the forest floor. Furthermore, those upstairs and downstairs forest denizens live closely linked lives. Soil-dwelling bacteria and fungi break down dead organic matter into molecules that aboveground plants use for food. Those plants, as well as animals, mature and die, leaving more organic matter to fuel the folks downstairs.

Like a well-insulated house, the soil protects its tenants from extreme temperatures and from rain and snow. It also provides a bulwark against predators that roam the surface world. But the dense, protecting soil also limits mobility. Soil creatures must be specially equipped in order to travel easily through their dark, constricting realm.

Earthworms and ants are the champion earth-movers, creating channels that allow air and water to enter the soil. While ants travel relatively far from their nests, earthworms work small areas, reprocessing vast amounts of soil into fertile "waste." In a single year, as much as thirty-six tons of soil may pass through the alimentary tracts of all the earthworms living in an acre of soil.

Of all soil-dwelling creatures, the most abundant are mites and springtails, insectlike creatures that literally eat their way through caverns of subterranean vegetation. The tiny, eight-legged mites lay their eggs on plant matter, which their larvae eat and convert into fresh soil. The bright-colored springtails are named for their ability to leap long distances during their search for decomposed plant matter to eat.

Both mites and springtails are prey to a host of soil-dwelling predators. They thus anchor one end of a food chain that extends to higher forms of forest "low-life," such as moles that feed on earthworms and shrews that eat beetles.

Those mammals, in turn, dig tunnels that function as underground byways for other subterranean species. Hibernating chipmunks, turtles, and salamanders sift and mix the soil when they burrow to winter dens. Cottontails and gray foxes excavate shallow dens as sanctuary from predators and harsh weather, while gray squirrels, hiding acorns for the lean season, further blend the earth. From microbes to people, thousands of species work the land upon which all life depends. [1]

INSECTS: AMAZINGLY ADAPTED AND ADAPTABLE

There are more insects than any other group of animals in the world. They live on top of mountains, in underground caves, in deserts and rivers, and in fields and forests. Their numbers are enormous, with over 900,000 species identified. There is great diversity within this large population, but all insects have the same basic structure, with each species adapted to meet the demands of its own particular environment.

Insects are arthropods—related to spiders, crabs, and lobsters—and like these cousins, they have jointed legs and an exterior skeleton. An insect's body consists of three main parts: head, thorax, and abdomen. The head contains the eyes, one pair of antennae, and mouthparts. The antennae are finely tuned sense organs, capable of feeling, tasting, smelling, detecting temperature, and receiving chemical stimuli. Some are long and slender, some quite feathery, and others clublike. The mouth of the insect is adapted to the food it eats. Thus a grasshopper's mouthparts are adapted for biting and chewing, a housefly's for lapping, a butterfly's for sucking, and a mosquito's for piercing and sucking. A few species, such as the mayfly, live as adults only long enough to mate and have no mouthparts with which to eat.

The middle section, or thorax, has three pairs of legs and usually two pairs of wings (sometimes one and occasionally none). For this reason, many muscles are located in the thorax. Insect musculature is highly specialized in many species—grasshoppers have about nine hundred muscles, compared to humans' eight hundred, and an ant can carry fifty times its own weight. Legs are adapted in as many ways as mouthparts. The grasshopper has hind legs specialized for jumping; the housefly has sticky pads on its feet, allowing it to walk up vertical walls; honeybees have specialized hairs on their hind legs that form "baskets" in which to carry pollen. The wings may be long, short, narrow, wide, leathery, or quite delicate, depending on the type and amount of flying the insect does.

The abdomen is divided into many segments and contains the heart, the digestive system, and the reproductive organs. On females of some species, the egg-laying device (the ovipositor) protrudes noticeably from the end of the abdomen.

Breathing is done through holes (spiracles) in the abdomen leading to the tubes that carry air throughout the body. (There are also some spiracles on the thorax.) Air is pumped in and out by the swelling and relaxing of the abdomen. The challenge of breathing under water has produced some remarkable adaptations, including the breathing tube of the mosquito larva, gills of immature caddis flies, damsel flies, and mayflies, and a system for carrying an air bubble used by certain water beetles.

For at least a third of a billion years, insects have been adapting to their environment. Of great importance to their flexibility in coping with harsh seasonal variations is the evolution of complete metamorphosis, resulting in the utilization of new habitats and food sources. The success of insects as a group is due to their having several major assets: flight, adaptability, external skeletons, small size, metamorphosis, and the ability to produce multiple offspring rapidly.

[2]

KATYDIDS

In a world of hungry predators, to be overlooked is half the battle. Let your enemy's jaws enclose a neighbor, while you go about the important business of being alive. For certain katydids of Central and South America, success is to be forgotten but not gone.

Katydids are grasshopperlike insects found throughout the world. Especially abundant in the rain forests of the American tropics, they represent an important food source for monkeys, bats, and birds. To avoid being eaten, many rely on mimicry. By night they roam the foliage in search of leaves to eat and other katydids with which to mate. By day they keep still, sitting on twigs or resting among dead leaves on the forest floor. Motionless, the katydids escape detection because they look like leaves.

One leaf-mimic in the Peruvian Amazon has brown twiglike legs and antennae, and plush green wings that taper into a convincing imitation of the drip tip found on many rain forest leaves. Another species has rounded wings that look as if they have been chewed by a caterpillar.

For almost every leaf in the rain forest, there seems to be a katydid copycat. But the insects do not mimic just pristine vegetation. The blotchy green, white, and brown pattern on the wings of one species suggests the splat of a bird dropping. Other specialists—pigmented with earthy beiges and browns—impersonate dead, dry vegetation; spots on their wings resemble the holes and decay marks normally found on fallen leaves.

Not all camouflaged katydids take after leaves. Some are patterned like tree bark. During the day they rest invisibly on trunks and branches. And some lichen-colored species perch only on branches with a bend in them. When the katydid settles into the angled corner of the limb, the overall appearance is of a smoothly continuous branch—providing no clue to lure a hungry bird or monkey.   [3]

CRICKETS

I find that an early-autumn evening would not be complete without the chirping of resident crickets. As the nights grow longer and cooler the music of these nocturnal creatures increases in intensity. They have the deadline of hard frosts to work against, for several successive nights of sub-freezing temperatures will bring their lives to an end. Before this occurs, mating (of which singing is a part) and egg laying take place.

The male black field cricket is responsible for much of the music we hear. With the edge of one wing rubbing against the opposite wing, he creates a chirping noise which is his means of claiming territory for himself and his mate. In the presence of a female field cricket, the series of chirps becomes much higher in pitch, and the male cricket moves rapidly about. Some of these sounds are made at the rate of seventeen thousand vibrations per second—hardly distinguishable to humans.

In many countries, especially China and Japan, crickets were so prized for their singing that they were kept in cages and carefully cared for. Actually since A.D. 960 in China, crickets have been kept for their fighting ability. Cricket fights were as popular then as horse racing is today. (They actually fed the crickets special diets, including mosquitoes fed on trainers' arms, and weighed the crickets in order to classify them!)

Crickets are considered by many people today as symbols of good luck, and are a welcome addition to household and hearth. They do add cheer to a silent home, but beware—they are voracious vegetarians, devouring everything from clothing to books!

The next time you look closely at a cricket, notice whether or not it has a long, spear-like ovipositor protruding from its abdomen. If it does, it's a female, for this appendage is an egg-laying device. (Both male and female crickets possess cerci—short barbs on their abdomens used to detect vibrations in the air and on the ground.) If you hear a cricket singing you immediately can assume it's a male, as females make no noise. [4]

MOSQUITOES

"Mosquitoes breed in standing or stagnant water," notes Dr. Gordon Nielsen, extension entomologist at the University of Vermont. "Eggs hatch in ten to fourteen days, with the new female able to reproduce almost immediately. Although the mosquito's natural life span is short, each female can breed several times, laying one hundred eggs at a time."

Over forty different species are found in Vermont, though generally only two to three species are active at the same time. Only the females "bite," gorging on blood from their victims to nourish their eggs.

"Mosquitoes home in on a target by using their antennae or sensors," the entomologist explains. "They find moist, warm skin, perspiration, and carbon dioxide in exhaled breath especially appealing. Dark clothing also helps them locate a target."

Insect repellents do not kill mosquitoes. They merely confuse their sensors so they're not sure if they should bite. Effectiveness is generally limited and reapplication necessary to ensure any degree of protection.

"The key to control is prevention," Nielsen stresses. "This means destroying breeding places before the mosquitoes have a chance to lay their eggs, or eggs already laid can hatch."

For mosquito prevention, dispose of tin cans, old tires, and other open containers around your home where rainwater may collect. Keep roof gutters cleaned out, and make sure house-cooler drains are properly connected to sewer lines or other drainage systems. Cover open ditches, and fill in low spots in your lawn.

Changing the water in birdbaths and fountains, encouraging dragonflies, and stocking ornamental ponds with mosquito-eating fish also will help keep down mosquito populations. So will emptying children's plastic swimming pools when not in use.

Nielsen adds that "bug zappers" generally are not effective in mosquito control as they tend to attract many large beetles and moths, but not many mosquitoes. Instead, if you plan to go out after dark when mosquitoes are most abundant, wear long pants and a long-sleeved shirt. Apply repellent to your clothing or face and hands as needed.   [5]

HOUSEFLIES

I was amazed by a housefly's ability to land on the ceiling and walk upside down. It's impossible to watch what actually happens when a fly executes this maneuver, but English photographer Stephen Dalton managed to capture the whole process on film. The fly senses when it's approaching the ceiling, reaches up to touch it with its front legs, flips the back part of its body forward in a half-somersault, and lands upside down on the ceiling. The fly can walk on the ceiling—the same way it can walk up walls and windowpanes—because it has sticky little hairs on its feet.

Unfortunately, in addition to helping houseflies defy gravity, the hairs sometimes accumulate pathogens—the types of microorganisms that cause diseases. Houseflies live in two different worlds—the indoor world of sugar bowls, plates of cookies left on the kitchen counter, and slices of bread waiting for someone to make sandwiches, and the outdoor world of manure, outhouses, compost heaps, and other decomposing matter. They lay their eggs and do a certain amount of feeding outdoors and then fly into the kitchen through an open door or window to crawl over whatever other foods they can find indoors.

Houseflies lay their eggs outdoors in warm, moist, decaying matter because that—and the microorganisms therein—is what their young like to eat. Within a day or so, the eggs hatch into legless white larvae called maggots that eat and grow enough to molt three times during the next four to five days. For the next stage of their life cycle, their final larval skin turns brown and toughens into a protective shell, inside which the insect pupates for another four to five days before emerging as a winged adult. The only way to discourage houseflies during these early stages of their lives is to cover or eliminate the decomposing matter they like to live in.

When the adults emerge as full-grown houseflies, ready to buzz into your house in search of a random meal, you can thwart them with screens. Any that do get in, you can swat with your trusty flyswatter. If you manage to kill a young female, you may be eliminating as many as a thousand offspring, and swatting older females and any males also reduces the local potential for reproduction.

Covering your food will discourage houseflies from spending time in your kitchen, and hanging sticky flypaper will offer them a lethal alternative. Yet another control that

I have discovered as a result of my own haphazard housekeeping are my resident spiders.   [6]

FIREFLIES

It is not true that we used to sit and watch the grass grow.
What we did in those energy-short days was sit and watch fireflies glow.

It was only in the evening after supper that we could sit. From dawn to dusk we had to be up and stirring. It was the Great Depression. Americans had not become oil addicts. Most stuff was done by muscle power. Television not only did not exist but also was unnecessary. By nightfall we were so weary that just being able to sit down was treat enough.

Fireflies now are increasing in southern New England after a few rough years during which pesticides commonly were drenched over the area repeatedly each summer. It was a worse time for bystanders than for pests in the insect world. In the last thirty years there have been entire summers during which I never saw a firefly.

It was during the firefly droughts that I most admired the Japanese, who have the good taste to release fireflies at their evening garden parties.

Fireflies always will be lightning bugs to most of us, because that was the name they bore when I was a child. We children admired the fireflies' helicopter-style flight. It enabled the insects to hang in space with their lanterns clear and flashing bright.

Most New Englanders know that the lighthouses along the coast beam different and distinctive series of flashes. In the darkness a mariner knows from the numerical pattern of flashes what light he is near and therefore his relative position. Most people, however, do not realize that fireflies also function through a code of flashes. Among fireflies the males are the dandies, flitting about and boasting about their maleness. The lady fireflies range from almost flightless to totally wingless. The male fireflies flash a distinctive series that identifies their species. The females generally live in the grass. When they read a message flashed by a male of their species, they flash back a here-I-am-big-boy. The female need not look over all the fireflies. The males of each species, when searching for females, fly at a characteristic altitude, some high and some low.

Fireflies are not the only animals that glow. Nor are they the only animals that transmit coded information through light patterns. Many deep-sea fish have impressive lighting arrangements, including what amount to searchlights in some species. The mystifying factor in all these lights produced by living organisms is that the lantern is always cold, or no warmer than the remainder of the body. The artificial lights produced by man's engineering all involve heat at some step.   [7]

HONEY

Bees gather nectar and pollen from the flowers, and the tiny drops of nectar secreted by your garden's flowers are the main ingredient in honey and the primary staple in the honeybee's diet.

In the search for food, the honeybee's body and legs catch and hold pollen grains. As the bee moves about from one flower to the next, some of the first plant's pollen grains are dusted off onto the stigma of the second and the cycle of cross-pollination is completed, allowing the fruit to form.

And what do you get out of this? Garden-fresh fruits and vegetables at the peak of their natural goodness, and, from the bee, honey . . . nature's finest natural sweetener.

The secret to beekeeping is in the modern beehive, resembling a multi-storied factory, where the parts are interchangeable for different styles of management.

The hive stand with its angled landing strip serves as the base of the hive, and the bottom board, with its adjustable entrance, rests on the hive stand and provides a floor for the hive.

The two lower stories of the hive are known as the brood chamber, where the queen bee lays her eggs and the baby bees are raised.

Bees cluster on the combs during winter and eat honey to keep warm. Before the first blossoms appear, the queen starts laying again and soon the worker bees are active in the field replenishing their supply of nectar and pollen. The colony continues to grow during the season, needing more room both for storing honey and for the rapidly growing number of worker bees.

The upper two stories of the colony are called supers and are equipped with beeswax combs. Beekeepers give their bees room for storing honey by adding these supers as the season progresses. The bees ripen the nectar into honey and cap the cells in the combs.

At the end of the nectar flow, the bees again prepare for winter and the beekeeper removes the surplus honey, leaving plenty of stores for winter use as well as for rearing a fresh supply of honeybees for the hive.

A colony of honeybees may contain from 7,000 to 70,000 bees. There will be one queen bee that lays all the eggs, a few drones (male bees) whose only job is to mate with new queens, and the rest of the bees will be worker bees (females) who gather the water, pollen, and nectar, clean and guard the hive, and feed the queen and growing brood.

In the active season, each worker bee during its 4–6-week life span produces only $\frac{1}{12}$ of a teaspoon of honey and about $\frac{1}{30}$ of a teaspoon of beeswax. Together, the worker bees of a hive may travel about 55,000 miles from June through August to gather the nectar to produce one pound of honey, a distance equal to about 11 round trips from New York to San Francisco.

It is because of this constant travel from flower to flower

that honeybees are so valuable for their pollination. A single hive of bees may produce from 100 to 200 pounds of honey a year, which would bring their total miles traveled to approximately 5,500,000 miles.   [8]

FEMME FATALE

Even its name sounds dangerous—black widow spider. But it's a good name for the dominating female of this poisonous spider species. For unless he is very careful in his courting procedures, the male spider may be eaten alive by an angry female black "widow."

The female black widow should be avoided by humans, too. For while her bite is seldom fatal to humans, it can be very painful, producing symptoms not unlike those of appendicitis.

Luckily, the female black widow would "much rather avoid people than bite them." Even when disturbed in her nest, she will usually try to escape rather than attack.

And she's also easy to recognize. The female, which can grow up to two inches in length (twice the size of the male), has a tiny yellow, or sometimes red, patch, often in the shape of an hourglass, on her abdomen. Her abdomen can swell to half an inch in diameter when it's full of eggs.

Found throughout the United States and other parts of the world, but mostly in warm climates, the female spins her sticky, untidy web in dark, dry places—under stones, in holes in the ground, around tree stumps, in log piles, garages, and basements.

In the United States and Canada, fatalities from wasp and bee stings far outnumber those from spider bites. While her venom sacs are small, the female's poison is up to fifteen times more potent than that of a rattlesnake. A black widow bite often causes nausea, swelling, and mild paralysis of the diaphragm, but most bite victims recover without serious complications.

The male black widow's poison is weak and ineffectual, so no wonder he's such a careful suitor. He approaches the female's web, and "taps out a kind of Morse code on its threads to find out if she's ready for him." If she's not, she may turn on him "in a fit of black widow anger. Or she may mate with him and eat him later."

More often, the female black widow, like most spiders, feasts on insects. When a victim gets caught in her web, she "wraps it tightly in silk, then swiftly bites it with two tiny fangs which deliver her potent poison."

Male black widows don't feed. In fact, they're very seldom seen at all. They spend most of their time wandering in search of females—which for them is a hazardous occupation. [9]

A WORLD OF SILKEN LINES

All spiders produce silk, and not only silk but several different kinds of it. A highly complex protein, it is synthesized in glands near the apex of the abdomen and issues forth through special spigots called spinnerets. When a spider moves from place to place it pays out a silk line. When it is ascending or descending, this silk line acts like the climbing rope of a rock or mountain climber. On a flat surface the spider lays out its dragline behind it. Male spiders can find females by following this telltale trace. Spiders also use silk for building webs, which are amazingly diverse in structure, as well as for making egg sacs, trapdoors, and the sperm webs of the males. The central role of silk and silk structures in the lives of many spiders is reflected in their superb sense of touch and exquisitely sensitive vibration receptors.

In a world of melodies and cacophonies played on silk, vision is almost always much less important. Most spiders have eight eyes arranged in two rows of four. They are generally much inferior to the eyes of insects and probably just movement detectors and monitors of light intensity. (Of course, some spiders have eyes capable of some kind of form vision and certainly of precise movement detection. The jumping spiders have relatively large eyes and use them to locate prey; even their courtship depends on visual signals.) The majority of spiders, however, have eyes that are small and that pale into insignificance in comparison with their vibration receptors. These spiders almost certainly live in a world where nearly all objects have a resonance or vibration fingerprint.

Spiders have become widely diverse in their ways of life; there are species whose habits are bizarre to the point of extreme improbability. The unifying theme among this diversity is that almost all spiders are hunters of one kind or another. The hunted, overwhelmingly, are the insects. Hunting techniques range from simple prowling to the use of sophisticated traps of marvelous construction. Spiders build complicated traps that rival those of man in apparent ingenuity. A review of webs and weblike structures would require an entire book, which has yet to be written, although it has been attempted several times. Discovery always outpaces the literature. [10]

DADDY LONGLEGS

Harvestmen have been with us since spring, but the young, which emerge from overwintering eggs, are very timid and usually hide under stones and other objects until they become fully grown in late summer, when they become more venturesome and come out into the open. It is because they are more commonly seen at this time of the year, the harvesting season, that they are called harvestmen. In Europe it has long been believed that a good harvest will follow their appearance in large numbers and that it is unlucky to kill one.

Many of us know these animals as daddy longlegs. They resemble spiders, with which they are often confused, but for some reason are not feared as much as spiders. Should you be unacquainted with these animals, they may readily be recognized by their long stiltlike legs and may easily be distinguished from spiders by the absence of a construction between the two parts of the body, the cephalothorax and abdomen, which are broadly joined to form a single unit.

As you watch the harvestmen run over the ground, you may well wonder how they can support their bodies on such long, thin, fragile legs and how they can move over the ground and through the grass without getting those legs caught in the grass blades. Eight legs would seem enough to carry a body not much larger than a grain of wheat, and there are times when they do get their legs entangled. But getting a leg caught is a trifling matter, for if the animal cannot extricate itself, it will merely "throw the leg off" and grow a new one.

The legs, of course, are much stronger than we would suspect and, strangely enough, separate quite easily from the body. This ability, combined with the legs' unusual length, has led some observers to believe that they serve as a sort of protective fence when the animal is attacked—that the enemy grasps a leg as the nearest thing to seize and is left with it as the harvestman "throws it off" and escapes.

Unlike their relatives, the spiders, the harvestmen do not have any silk glands and hence cannot spin webs or retreats. They feed, for the most part, on dead insects but are also known to kill small ones for food and to suck juices from various soft fruits and vegetables.

Before the first frost, the females lay their eggs in the ground, under stones, and in the crevices of wood. Unlike the eggs of spiders, they do not hatch until the following spring. In the north the winter cold kills the adults with one exception: one northern species lives through the winter as adults. In the south most harvestmen hibernate in some safe retreat—for example, beneath rubbish.   [11]

POND GIANTS

When we think of predatory animals, it seems that the first examples which come to mind are usually the larger mammals and birds of prey. Yet, probably the most numerous predators are the insects.

The first insects may have been scavengers, but at an early date developed the ability to feed on plants. The hunters of the insect world evolved from these early insects in spite of the abundance of plant foods. This is probably due, in part, to the high protein content of a carnivorous diet. The food has already been converted from plant to animal tissue.

In nearly all groups of insects where the predatory habit is well-established and widespread, there are striking anatomical modifications which assist in the capture and handling of prey. The front legs are often exquisitely adapted for seizing and frequently have curved, sharp claws.

The people at Vermont Institute of Natural Science tell us that the most powerful and fascinating of all the predaceous insects is the giant water bug. Among different species the individuals vary in size, with the average being about two inches long. There are some tropical giants which may be more than four inches in length, however.

The giant water bug's body is flat and oval with a keel along the underside. This design makes the insect well-adapted to aquatic locomotion. It has sharp-hooked front legs for grasping and two pairs of ciliated, oarlike hind legs for propelling it through the water.

This hunter is quite skilled at its livelihood. Once the prey is captured with the grasping legs, a venom is injected with a short, powerful beak. This quickly kills the victim and dissolves the tissues. The body fluids are then siphoned out by the sharp beak. The giant water bug is thus capable of subduing relatively large animals, such as small snakes, frogs, and fish.

The giant water bug usually lurks on muddy bottoms of lakes and ponds, often covering itself with mud or leaves. Occasionally, the water bug will leave the watery environment and fly.

One of the most interesting behaviors of the giant water bug is that the females in two of the genera glue their eggs to the backs of the males, who carry them around until they hatch. It's said that the males don't take kindly to this, but can't do anything about it. Specimens thus bearing eggs are especially fascinating.

Predators live by strength and skill, while prey continue to exist through the use of effective camouflage or reproductive capacities. Unseen little dramas of nature are being enacted every minute of the day. The aquatic world beneath the pond surface is no exception. [12]

A THREAT OUTDOORS

They've been spotted in gardens, under kitchen sinks, on houseplants, and even in rural mailboxes. They look ferocious to the unknowing, but to others they're nothing but an annoying home and garden pest. This perennial summer visitor is the European earwig.

"Although less than an inch in size, the earwig has a frightening appearance and an offensive odor," says Dr. George MacCollom, extension entomologist at the University of Vermont. "It has a reddish-brown color with a pair of pincers at the base of its abdomen. It does not sting or bite but may try to pinch when touched."

Earwigs are annoying but relatively harmless in the house. However, they do have a disconcerting habit of appearing suddenly when their hiding place is disturbed. They favor damp, dark places and often hide in cracks, hollow aluminum doors, or the pantry. In the garden it's a different story. They can be extremely destructive to a number of flower varieties, including asters, carnations, marigolds, zinnias, and roses, and to vegetables such as beans, carrots, corn, lettuce, and potatoes. They also may attack peaches, raspberries, apricots, and other fruits.

"The young generally appear in July and August and will start searching for food when they are a few weeks old," MacCollom points out. "Earwigs tend to scavenge at night and hide during the day, so you may not always be aware of their presence, especially in the garden." For the most effective control, insecticides must be applied during the warm, dry weather in early summer, although later treatments will help reduce populations. Always use only those products registered for use in your particular garden situation. Remember never to water treated areas for the first two nights after treatment.

MacCollom adds that control in populated areas will be successful only when everyone in the neighborhood treats their gardens. Earwigs tend to roam freely and may move back into your garden from another, even though you spray regularly.

"For indoor control, spray along baseboards, drainpipes, and under carpeting with a product recommended for indoor use," the entomologist says. "If you don't already have earwigs inside your home, take care not to bring them in. Inspect firewood, plants, lawn furniture, barbecue grills, and other outdoor equipment before bringing these items indoors."

Traps consisting of several boards banded together with a pencil or stick between each board also may provide good control, especially near houses. Place near plantings and foundations to attract earwigs. To kill these pests, remove the pencils or sticks, thus crushing them between the boards.

[13]

A HOMESPUN REMEDY FOR FLEAS

Knowing a thing or two about fleas' habits may help you outwit them.

You know you're in trouble when: 1. Certain areas of your carpet seem to hop. 2. Sitting on the couch becomes an experience in acupuncture. 3. Your dog and cat act like animal contortionists in their efforts to scratch and bite six itches at once.

In other words, your house has become a giant fleabag. What's worse, commercial insecticides are often not only smelly and poisonous to other life, but ineffective. Fleas, highly specialized bloodsucking parasites with the evolutionary hardiness of cockroaches, have become resistant to the pesticides commonly found in insect bombs and flea collars. In fact, many veterinarians advise pet owners not to waste their money on flea collars.

But there's more than one way to flummox a flea. In my copy of the *Old Farmer's Almanac*, I find a few suggestions:

- Vacuum rugs and upholstered furniture daily to remove fleas, eggs, and larvae. Fleas spend most of their time away from their hosts and can survive for several weeks without feeding, so keep up the vacuuming diligently for at least a month. Be sure to block up any exits from the vacuum cleaner after vacuuming lest the fleas hop back out. Vacuum your pets if they will let you (some love it; some would sooner die a horrible death).
- Add minced fresh garlic and a sprinkling of brewer's yeast to your pet's food to repel fleas.

Your veterinarian can supply you with a new treatment for fleas, a non-aerosol spray for pets (Sectrol is one brand name) based on the chemical pyrethrin, which is extracted from the dried heads of certain varieties of chrysanthemums. Another new veterinary product that is especially safe for indoor use is a rug and upholstery non-aerosol spray called Duratrol that microencapsulates tiny drops of an effective insecticide.

Many pets are made miserable not only by the flea bites themselves, but also by an allergic reaction to flea saliva, a persistent condition that is often treated with cortisone-related medication.

For most effective control of fleas, start early in the spring and be diligent. With any luck, your fleas will flee.

A Few Facts About the Infamous Flea
- Fleas can jump 150 times their own length, vertically or horizontally. This is equivalent to a human jumping 1,000 feet.
- Flea bodies can withstand tremendous pressure, their secret to surviving the scratches and bitings of the flea-ridden host.
- Fleas are covered with bristles and spines that point backward—that is why it is so difficult to pick a flea from your pet's fur.
- As carriers of plague, transmitted from infected rats via fleas to man, fleas have killed more people than all the wars ever fought. [14]

SNOW FLEAS

In late winter, you may have noticed large numbers of minute, wingless insects jumping about on the snow surface. Commonly known as snow fleas, these insects belong to the order Collembola, or springtails.

I noticed them around our house and also when I was working in my woodlot. I checked with the Vermont Institute of Natural Science and they told me that there are some two thousand species of springtails in the world and they can be found on every continent, including Antarctica. They occur in moist vegetation, forest litter, under bark, in decaying wood, and on water surfaces. They belong to one of the earth's most beneficial janitorial unions, the scavengers, feeding on decaying matter such as leaf litter and thereby playing an important role in the formation of humus.

Most species of springtails possess a long slender tail-like structure, or furcula, at the posterior end of the abdomen. The tip of the furcula is held close to the under-surface of the abdomen by a clasp-like structure. When the clasp is released, the furcula snaps against the ground and propels the insect forward and upward.

Research seems to indicate snow fleas and other spring-tails swarm in response to population pressures. Under ideal conditions of food and temperature when predators are scarce, the population rapidly increases, creating a food shortage. On a suitable day, when temperature and humidity are right, the insects migrate in great numbers to the surface of the snow, emerging around the tree trunks, rocks, and weed stems where snow has partially melted away. Great numbers of them will die within twenty-four hours, while a much smaller number will find their way back down to the leaf litter to survive the rest of the winter. There is still much to be learned about springtails before we can be sure of the reasons for the late-winter appearance of these insects.

[15]

THE LIFE CYCLE OF A MOTH

The notorious clothes moth is the only moth many people know, but ten thousand other species are indigenous to the United States and Canada. Some are overlooked because of their nocturnal habits; others are frequently taken for butterflies.

Moths do resemble their diurnal cousins. Like them they wander about on four fluttering wings that derive their color and pattern from almost microscopic overlapping scales. Like butterflies, most moths have a tubular sucking organ called a proboscis.

On the other hand, there are distinct differences between the two *lepidoptera*. Moths usually have hairier, more

robust bodies than butterflies, and antennae that are feather-
or awl-shaped, but never knob-tipped like the butterflies'.
Instead of resting with their wings held vertically above their
backs as butterflies do, moths spread theirs flat or hold them
back like a miniature pup tent. Generally, if you see something at night, it's a moth; in the daytime, a butterfly.

Moths become moths by passing through four stages
of complete metamorphosis—egg, larva, pupa, adult. The
female gets the cycle started with her enticing scent, which
some males can detect over two miles away! She lays her
fertilized eggs on food plants or the ground.

The larvae usually hatch within a few days. Some are
tiny enough when full-grown to burrow between the upper
and lower surfaces of a leaf. Others, like the seven-inch
hickory horned devil, are huge caterpillars. At this stage they
are vulnerable to birds and many predacious or parasitic
insects.

In the pupal stage, wings begin to form, chewing mouth
parts change into proboscises, and organs of reproduction
develop. Some larvae spin silk cocoons for protection during
this stage, each species making a distinct type.

Entering its final stage, the insect crawls from the cocoon, not quite a moth. Its abdomen is still elongated and
the wings-to-be are mere crumpled stubs. Then the body
pulsates and the stubs tremble as fluid is pumped through
their veins. Within minutes these stubs unfurl and expand,
until at last they are wings.

One of the most distinctive moths is the narrow-winged
sphinx often seen on flowers in the early evening. Notewor-
thy, too, are the tiger moths, many of which are spotted and
streaked with yellow, black, orange, pink, brown, and white.

But for sheer elegance, the luna moth reigns supreme.
It is a large furry creature, with sweeping, swallow-tailed
wings of the most delicate pale green hue. Each purple-edged
wing has a "windowpane" surrounded by a lovely gold and
black eyespot.   [16]

WHY DO WOOLLY BEARS CROSS THE ROAD?

When asked this question recently, I couldn't resist the classic response: "To get to the other side." And that response is true, but let's look further behind this behavior of the woolly bear caterpillar. These insects belong to the tiger moth family. The larval and caterpillar stage of these moths generally are quite large, active, and covered with dense, long hairs. In fact, most of the larvae of different members of this family are sometimes called woolly bears because they are so furry.

But the woolly bears—with the brown middle stripe and black at either end—are a familiar sight to us all, especially at this time of year, when they are so often seen wandering across roadways and paths. Known to wander for a particularly long time in search of a well-protected spot for the winter, these caterpillars are also the speed demons of the larval world. Woolly bears can walk up to four feet per minute, or one-half mile per hour. That is very fast for a caterpillar, so they can cover a lot of ground looking for a place which will provide adequate protection from predators as well as sudden temperature changes. They are most often found hibernating under loose bark, logs, or rocks.

Among insects, it is somewhat unusual that woolly bears overwinter in their larval stage. When spring arrives, the caterpillar awakes, resumes eating grasses, plantains, or other low-growing plants, and shortly thereafter pupates. While spinning its cocoon, the larva loses some of its hairs, which, with the silk, become part of the oval brownish cocoon. The pupal stage lasts about two weeks, after which the nondescript, brownish adult emerges in early summer. This light brown moth is one and a half to two inches in size and usually has a few small dark spots on its wings. This is the Isabella tiger moth.

Shortly after hatching, the females mate and lay egg clusters on various food plants. These eggs hatch in four or five days. The new larvae feed together for several days but then disperse, and by three to four weeks they have undergone six molts and are ready to pupate. The adults which hatch from these pupae lay more eggs, and it is the larvae from this second generation which will then be seen in the autumn looking for wintering quarters.

Isabella tiger moths and their woolly bear caterpillars are found throughout the United States. The larvae can be picked up gently without harming either human or insect. Their method of rolling up into a tight ball is a wonderful protection from predators, who are presented with a mouthful of fur. As well, the little roly-poly balls seem almost slippery and very hard to hold on to.

These miniature bears are a friendly sight to us as they try "to get to the other side" and are well worth a gentle, close-up look. However, I would not recommend formulating any significant decisions which depend upon the severity of the winter using the relative size of the woolly bear's stripes. From all accounts, this is unreliable at best!   [17]

TIGER SWALLOWTAIL BUTTERFLY

That yellow-and-black object flapping in the treetop probably is a tiger swallowtail butterfly. It could be a piece torn from a tangled kite or even a goldfinch male, but in mid-July the chances are that yellow and black in flight signifies the presence of this oversized butterfly. The tiger swallowtail has a five-inch wingspread, which is a fifth larger than the black swallowtail wingspread and double that of most butterflies.

The prevalence of tiger swallowtails in mid-July can be explained by the overlap of a second brood of the summer hatching while a few members of the first brood are still a-wing. The swallowtail has only two broods in southern New England, and the mid-July recruits will be the last group that flies this year. They will produce next spring's first generation, which will remain through the winter as chrysalids. Chrysalids are the final form of the insects before they emerge as butterflies.

The tiger swallowtail breeds from Alaska to the southeastern United States, one of the broader ranges among butterflies. It has one brood annually in the northern areas, primarily in Canada; two broods in southern New England; and three broods farther south.

The tiger swallowtail female resembles the male in northern areas. To the south of New England, the females in two-brood populations occasionally are black, and farther south in the three-brood populations, are usually black.

The great butterfly expert, Samuel Hubbard Scudder of Cambridge, Massachusetts, wrote of the tiger swallowtail:

The flight of all our swallowtails is nearly the same, but this butterfly is perhaps peculiar for its sailing and soaring; it flies high and low, rising and falling alternately several feet at a time, usually moving at about 20 feet from the ground; it flutters in and out among the branches of a tree, from base to crown.

The report not only furnishes an adequate description of the tiger swallowtail flight pattern but also places it where one might most expect the butterfly—among trees.

The tiger swallowtail enjoys a wide distribution, partly because it is less selective regarding the food for its caterpillar. While it chooses trees as the site for laying eggs, it will use a broad spectrum of species, including cherry, birch, mountain ash, poplar, willow, tulip tree, ash, basswood, maple, and magnolia. Many butterflies refuse to lay eggs on anything except a single plant species.   [18]

HOW TO PLANT
A BUTTERFLY GARDEN

Robert Frost called them "flowers that fly."

The lives of butterflies are rooted in the earth. If one understands their needs and their habits, butterflies can be cultivated as one might cultivate flowers . . . and with the same prospects of success.

Once butterflies move into a favored habitat, they seldom stray very far. Some species, especially the smaller ones, are so territorial that they will defend their chosen areas against all intruders, including birds, cats, dogs, people, and Frisbees. Angry butterflies have pursued trespassers for five hundred yards; but the nice thing about being attacked by a butterfly is that one rarely knows it. There is seldom physical contact.

The concept of a butterfly garden is not new. Winston Churchill's butterfly garden was among his more consuming and less criticized passions. He would even cheat by stocking it with choice specimens just before the arrival of dignitaries whom he wished to impress.

One cultivates butterflies by first cultivating the plants that they feed upon. Even without the butterflies a butterfly garden can have great aesthetic appeal when assessed by conventional standards. But the first requisite in successful butterfly cultivation is reordering one's concepts of what is "beautiful" and what is "proper."

Take, for example, the sacrosanct tradition of the American lawn. We agonize over our lawns: pulling, plucking, trimming, cutting, pouring millions of pounds of chemicals into them each year—all in an effort to maintain them in unnatural and vulnerable monocultures. The importance that we attach to lawns is such that the condition of one's grass is frequently thought to reflect the condition of one's character. Neat, enterprising Americans simply do not dwell amidst disheveled hayfields. Yet few environments are more sterile than a well-manicured lawn, and their proliferation is one reason that butterflies are fading from the American scene.

One need not let one's lawn go to seed in order to attract butterflies. But perhaps there is a hard-to-get-at patch or, even better, a strip that will, if you allow it to revert, save you work. For those who are nagged by spouse or conscience to mow the lawn, butterflies are a fine excuse to continue reading the paper. Butterflies depend most on the very plants lawn-lovers work hardest to eliminate, such as dandelions, clover, thistle, nettle, goldenrod, and ragweed. It's a good idea to cut back some of these plants as the season progresses so that new generations of butterflies—in both adult and larval stages—will have tender new food sources.   [19]

WATER &
AQUATIC LIFE

WATER SUPPLY:
ROMAN AND MODERN

From Spain to Turkey, in North Africa and especially in Italy, the remains of Roman aqueducts still march across the countryside, graceful reminders that the problems of water supply weren't invented in the twentieth century.

Sextus Julius Frontinus, who was appointed *curator aquarum*, or water commissioner, for the city of Rome in A.D. 97, proudly boasted of the practical beauty of the Roman aqueducts compared to the useless monuments of Egypt and Greece. In his two books on the Roman water supply, translated into English in 1896 by Clemens Hershel and reprinted in 1973 by the New England Waterworks Association, Frontinus gives a very complete picture of Roman water-supply problems.

The Romans faced some of the same troubles we face today: getting the water from where it is plentiful to where it is needed, accounting for water system losses, repairing and maintaining old, leaky water systems, metering the amount of water used, and even problems of pollution, particularly with lead pipes. They also used tile and clay pipes, and, of course, the famous stone aqueducts.

When Frontinus became water commissioner, he made a hydraulic survey of the entire system. He learned that though the supply was plentiful at the source, much of the water was lost before it reached its destination. To solve the problem, he repaired the aqueducts, set up a regular maintenance schedule, and checked for illegal connections to the system and faulty or tampered-with metering devices.

The Roman "meter" was simply a piece of lead pipe of a particular diameter, stamped with the imperial seal to indicate that the user had the emperor's permission to take water out through a hole of that size. Water could flow out of the hole continually. Obviously it was easy to have "metering" problems, accidentally or intentionally.

The modern world is still trying to cope with the intertwined problems of water quality, water supply, and water conservation. Modern industrial development has brought us pollution problems on a scale which the Roman world never had to face, but on the other hand we also have ways to tackle them that Sextus Julius Frontinus never dreamed of. Frontinus was eminently qualified for the job, but he just didn't have the tools we have.

When Frontinus did his field work, he had to bring along a retinue of eight or ten mathematicians to figure the amount of water going into and out of the system, using huge tablets and the Roman numeral system, which not only was cumbersome in itself but also lacked the concept of zero. Multiplication and division were a matter of sequential adding and subtracting . . . in V's and X's, L's, and C's. Tiny counting pebbles called calculi helped some, but they were far from being calculators in the modern sense of the word.

Frontinus had some concept of water pressure and realized that even if the holes were the same size, more water came out of a hole that was near the sea than out of one in the mountains, but he had no way to figure out the difference. What Frontinus would have given for a simple hand calculator instead of a pile of pebbles—not to speak of algebra, a few basic laws of physics, or a computer.   [1]

PLUMBING: A LOOK BACK

Despite painfully slow advances in sanitary practices (and sometimes their disappearance altogether), plumbing has been with us throughout written history. In imperial Rome, for example, elegant communal baths accommodated as many as three thousand people. However, bathing was not necessarily for getting clean. The baths were social gathering places, suitable for conversation, relaxation, and who knows what else. In Rome the sexes remained segregated. In the baths of the Ottoman Empire they did not. These Turkish baths, in fact, contained small private rooms called bordellos, where, it seems, most behaviors were permissible.

While there may not be a connection, such "friendly" bathing habits immediately preceded the Middle Ages, Europe's darkest years. Plagues swept across the continent, wiping out 25 million people—a quarter of the population. Sanitation simply did not exist. Among the few great achievements of the Middle Ages were the remarkable castles built throughout Europe. They were small fortresses, cities under a single roof. Some, not so small, contained as many as 1,500 rooms. Their defense, as we all learned in grade-school history class, relied in part on the moats that surrounded them. But alas, that explanation is not entirely true. In fact, the moats did provide effective protection from invading enemies, but not by design. The castles contained no bathrooms. They did, however, have privies built into the outside walls that were dumped directly into the moats. The moats were nothing more than stagnant cesspools that must have been incomprehensibly disgusting. Only a fool would have crossed through one.

Medieval moats highlight the difficulties accompanying waste disposal. In seventeenth-century England, the problem reached staggering proportions. In 1609 London built a water system that brought clean water from a distance of forty miles. With plenty of available water, the use of the water closet flourished. Then came the Industrial Revolution. Millions of people moved from rural areas to the city for work. In 1778 Joseph Brahma (and not, as legend would have it, John Crapper) received a patent for the float-and-valve flushing system still in use today. While water closets became more common, they were connected to cesspools by unventilated pipes. Not only did these WCs stink to high heaven, but they were also serious sources of bacteria and infection. In the true spirit of treating the symptom instead of the disease, the Stink Trap was patented in 1782. It successfully eliminated the smell but did nothing to stop the spread of disease.

Society simply did not know how to deal with the problems created by industrialization. In 1847 the British Parliament created a sewer commission and required that every house have some sort of sanitation: an ash pit, a privy, or a water closet. In 1848 they passed the National Public Health Act, a model plumbing code that much of the world has followed. But changes were slow in coming. Between 1849 and 1854, twenty thousand Londoners died of cholera. The Thames—the source of most of London's drinking water— was also its sewer! The city had a population of 3 million and no waste treatment. All of the city's human and industrial waste flowed into the Thames River.

In 1859 Parliament actually had to be suspended for a short time because of the unbearable stench. In 1861 Prince Albert died from typhoid. In 1871 the Prince of Wales almost died from the same disease. Moved by his illness, his recovery, and the related sanitary conditions, he reportedly said that were he not a prince, he'd like to be a plumber. From that point on, sanitation became a public concern, but it was hard to change old habits and fears.

For four thousand years the place where a person goes to do his or her "business" has never had a straightforward name. The bath or bathroom, after all, is not just for taking a bath, and the "necessary room" is a little oblique. The Israelis went to the "house of honor," the Egyptians to the "house of the morning," Romans to the "necessarium," and Tudors went to the "privy" or the "house of privacy" (or they went to the "Jakes"—Jack's place—because everyone had to go—now known as the "John"). Even sailors, a hardy lot with a reputation for direct language, go to the "head." Another favorite is the "loo." That word may have entered the vocabulary in this way: whenever a Frenchman tossed a load from his window, he first hollered, "*Guardez l'eau*"— watch out for the water. It was shortened to *l'eau* and soon became loo. [2]

SMALL WETLANDS

Across the country small wetlands vanish by the thousands every year. These miniature wetlands, ranging from a few square feet to perhaps ten acres in size, seem insignificant. Often poorly protected by law, they are easily dismissed, drained, or filled, then planted or built upon and soon forgotten. Their wild inhabitants will then have reached the end of the line.

Biologists have long known that the small wetlands— whether they hold open water the year round or not—are big wildlife producers. In parts of the prairie "duck belt" of Minnesota and the Dakotas, nearly half of the potholes dotting the countryside cover less than a quarter-acre each, and nearly one-fifth of them are dry for part of the year.

Yet one survey showed that 75 percent of the prairie country's breeding ducks, as well as 96 percent of its breeding shorebirds, are produced on these seasonal ponds. Invertebrates produced here feed the female and young blue-winged teal, pintails, mallards, gadwalls, and shovelers. Elsewhere, too, small wet areas are prime wildlife producers. One study listed seven species of amphibians as well as abundant reptiles, birds, and mammals in the small wooded wetlands of Maine. Some species are totally dependent on these miniature wet areas, which also become choice hunting grounds for hawks, owls, herons, and other predators.

A Massachusetts study revealed that, in spring and summer, black bears spend fully 60 percent of their time feeding in the small forested wetlands. White-tailed deer use them extensively. Among others, these wetlands are home to wood ducks, black ducks, mallards, yellow-throats, northern water thrushes, yellow warblers, and great-crested flycatchers.

Furthermore, even the small wetlands that appear to be isolated can be part of a larger wetland mosaic. And, as one Fish and Wildlife Service publication sums it up, "All of the wetlands are degraded if parts of the complex are destroyed."

Consulting biologist Robin Hart, who recently completed a wetlands study for the Florida Game and Fresh

Water Fish Commission, says that agencies ignore small isolated wetlands because they are not parts of identified surface-water systems: "But these deserve protection just as much as the larger ones do. Small size does not indicate that they are of less importance as wildlife habitat. Such areas can be of tremendous importance to waterfowl, wading birds, and amphibians, and as feeding areas for upland mammals, raptors, and passerines." [3]

THE ESTUARY

The rockbound New England coast has a soft and vulnerable underbelly. Here and there, where its hardened edge relents to let water drain from land to sea, a rich biological stew cooks. Most of us eat regularly from this dish, sharing it with thousands of smaller marine consumers.

Technically speaking, an estuary is the merging of fresh and salt waters. Its characteristic features include salt marshes, tidal flats, sand flats, and sandy beaches. Certain species of plant and animal life have evolved to live under the frequently harsh conditions that are a common element of the estuarian environment—changing salinity, temperature, and oxygen supply.

In spite of the adversities, biologists estimate that an estuary is among the most productive ecosystems on earth. It is considered ten times as rich as adjoining coastal waters, themselves about ten times as productive as the open ocean.

No single sum can account for the entire productivity of these unique ecosystems. For instance, it is estimated that 70 percent of commercial fish species landed between Cape Cod and Canada come to estuaries to feed or breed. A monetary figure may be attached to the catch, as well as to the shellfish and baitworm harvests that are also largely dependent on coastal wetlands, but the food web that supports these populations cannot be so easily quantified or delimited.

Estuaries are known to play a significant role in the recycling of nitrogen and sulfur into forms usable by primary producers. Some estuaries actually export nutrients into coastal waters, supporting populations that live several miles down the coast. Estuaries also serve as a buffer against severe storms; they rebuild themselves more easily then rigid coastlines that have been damaged by wave action.

What makes an estuary so productive? The answer has to do with the nature of energy flow and the abundance of plant and animal life. Twice daily, tides sweep into the estuary from coastal waters, circulating and depositing nutrients and removing waste products. Because of this input, energy that organisms would otherwise spend on maintaining their existence is applied to growth. Tidal pools in the marsh are refilled and replenished each day, allowing organisms to thrive there under tolerable conditions.

Furthermore, many estuarine organisms are small. Zooplankton, tiny crustaceans, and other microorganisms live and die in rapid succession, releasing their nutrients quickly into the environment. As a result, nutrients do not stay tied up in biomass for long periods. In addition, the muddy lower portions of an estuary support a great diversity of primary food types. Microalgae, phytoplankton, and detritus are easily obtained by the worms, clams, and dozens of filter feeders that live in the few centimeters of mud just below the surface. There, somewhat buffered against the drastic swings in heat and oxygen conditions above them, they live in huge numbers. [4]

ANOTHER WORLD, THE CORAL REEF

Coral reefs flourish throughout the world in shallow, well-lighted seawater where temperatures exceed 70 degrees Fahrenheit throughout the year.

Those fortunate enough to be able to travel to the tropics can find a world of color and patterns which can stop one's breath. The reefs are made up of heads of coral polyps. Each polyp is an individual animal which secretes limestone in the form of a protective cup about its body. The animal uses tiny tentacles to capture its food of plankton. Additional nourishment is obtained from microscopic algae which live embedded in the tissues of the coral polyps. These tiny plants add their color to the coral itself. The varying colors, shapes, and textures of the different species of coral create a huge flower garden appearance on the sea floor.

Even more striking than the coral are the tropical fish they attract. Their colors are vivid: purple against bright yellow, black on silver, iridescent aqua against royal blue, and other startling contrasts displayed in such patterns as spots, stripes, bands, and checks. Names like jackfish, big-eye, trumpetfish, butterflyfish, parrotfish, and balloonfish suggest the variety of interesting shapes of tropical fish.

The behavior of the reef fish is a treat to behold. The parrotfish uses its beak, which actually is fused teeth, to crush away at the coral reef, ingesting polyps, algae, mollusks, and bits of limestone.

The damselfish, delicate of name but not of dining habits, attacks algae, small crustaceans, and other fish with lightning-like charges. Even the tiniest of this group are extremely territorial and will frighten off much larger fish.

Goatfish use appendages located under their chins to stir up the sandy bottom of the reef. Small crustaceans are then exposed and gobbled up. Other freeloading fish follow the goatfish in hopes of an easily obtained meal.

Occasionally a barracuda will swim slowly by. It will move swiftly when it finds food to attack with razor-sharp teeth. It will not molest a human swimmer unless it mistakes a sudden movement or a flash of jewelry for a fish. Nevertheless the coral reef does have its dangers: electric rays, scorpionfish, stinging coral, and jellyfish. The informed and careful snorkeler will avoid these creatures while exploring the exciting and beautiful world of the reef.   [5]

TSUNAMI:
WHEN THE SEA QUAKES

Unlike ordinary wind-generated surface waves, tsunami are true gravity waves with exceptionally long lengths and periods. The tsunami is not a single wave, but a series or "train" of giant oscillations, similar to but incomparably larger than ripples radiating out from a pebble dropped in the water. More than 150 miles can separate two successive crests, which in open sea may average only two or three feet in height. With a slope this gradual, the waves go unfelt by passing ships.

Coastal topography determines how the tsunami makes landfall. If deep water extends close inshore, the waves' impact will be minimal, but will exceed high-tide levels like a vast surging flood. However, if the sea floor slopes gradually up to the coast, the great waves begin to "feel the bottom." This bottom friction shows their forward momentum, causing a "bunching up" as successive waves are pushed closer together. The "killer" in a tsunami train is often somewhere between the third and eighth waves, as displaced kinetic energy is transferred upward into towering walls of water, sometimes exceeding one hundred feet in height. The worst possible coastal scenario is the V-shaped harbor, where the force of an incoming tsunami is concentrated and funneled into a sea wave of monstrous proportions.

The first sign of an incoming tsunami train is often an ominous withdrawal of the sea from the shore—a massive outflow of water far beyond the limits of dead low tide—which occurs when a seismic trough, rather than a crest, reaches land. As the water recedes, a cacophony of hissing, rattling, sucking, and boiling sounds can be heard as pebbles, rocks, shells, and assorted debris are drawn out to sea. Whole bays empty in minutes, stranding myriad forms of marine life and exposing long-sunken wrecks to view. In the past, unknowing onlookers rushed out gleefully to examine these unexpected treasures. Few if any ever survived their folly, for minutes later the sea returned in a tremendous surge of churning water too fast to outrun.

Seismically generated waves attain astounding speeds while transiting an ocean. A tsunami's forward speed is directly proportional to the varying depth of water through which it passes—the deeper the water, the greater the speed. In water 20,000 feet deep, a tsunami shock wave will race along at 545 miles per hour. Moreover, depending on the original disturbance, tsunami can cover thousands of miles before finally dissipating. The world's longest tsunami waves come from Chile and can travel more than 12,000 miles across the Pacific—nearly half the world's circumference—to break against the shores of eastern Siberia.   [6]

INTERTIDAL CREATURES

If you find seaweeds washed up on the shore, look carefully through the holdfast; you may find very small starfish, shrimp, clams, or other creatures seeking protection there from the waves and predators.

It is vital for intertidal creatures, which are exposed to the drying effects of ocean breezes, beating sun, and fluctuating temperatures, to have means for retaining moisture. Periwinkles, a small snail species, have a trapdoor called an operculum that closes moisture within the shell. Limpets adhere to rocks using a flat, muscular foot around which they press their shell to effectively trap water between the foot and the shell's margin. Barnacles have doors on top of their conical shells that securely seal in water until the high tide returns.

The carnivorous sea creatures of the intertidal zone are remarkable predators. A starfish typically feeds on clams or mussels by wrapping its arms around its victim and pulling the shell apart with its tubed feet. By protruding its stomach through its mouth and into its victim's shell, it digests the soft body parts of its victim externally. Afterward, all that remains is an empty shell. Sea anemones, though sedentary, have tentacles with sting cells that paralyze living prey. The tentacles then draw the prey into the anemone's mouth. Dog whelks, which prey upon barnacles, blue mussels, and occasionally other dog whelks, actually drill holes into their victims' shells with a structure known as a proboscis, and then consume the soft parts within. Voracious feeders, dog whelks can markedly alter species composition in an intertidal zone.

Herbivores along the sea's edge feed primarily on various algal species. Limpets, periwinkles, and chitons scrape blue-green algae from the rocks with file-like structures called radulae. Sea urchins feed on encrusting pink coraline algae using beak-like jaws containing five teeth.

The delicate devices used by filter feeders are just as fascinating, although often harder to observe. These shore animals are generally sedentary, straining plankton brought in with each tide. Barnacles are the most evident filter feeders along the rocky shore. While underwater, their cover doors open to reveal six pairs of featherlike appendages. The feathered net sweeps rhythmically in and out of the door, catching plankton. Mussels and clams funnel water through siphon tubes into their shells where they can filter out food. Although they may resemble plants, sponges are actually colonies of animals that continually filter seawater for nutrients and tiny organisms.   [7]

THE POND

Ponds are places of magic and mystery. They have an aura that draws an inquisitive mind like a magnet. Suppose someone asked you to define the word *pond*. You might say, "Well, a pond is a place where the bottom is very muddy and where many different kinds of plants grow, like cattails and water lilies. Lots of things make their home in a pond: frogs, turtles, dragonflies, snakes, and birds are just a few. And there are many kinds of newts and other creatures hardly bigger than a speck of pepper scurrying around in the water."

This answer touches on the major characteristics of a pond, and it considers a pond to be a distinct ecological unit with certain conditions and boundaries. A pond is a shallow body of water with a muddy or silty bottom that generally supports aquatic plant growth from shore to shore. Plants usually ring the shoreline. Because of the shallow water found in a pond, its temperature changes significantly each day during the warm season. A temporary warm layer forms on top in the summertime, which is why a swimmer dives to the bottom to seek cold water. This warmer surface layer often disappears by the time dawn arrives on a cool morning, especially if there is a breeze churning the water. There is also a great variation in dissolved oxygen levels during each daily cycle. The abundance of living things and decomposition occurring in a pond account for its high levels of respiration and carbon dioxide.

A special kind of small pond, called a vernal pond, is ephemeral, existing only during the wet spring period and drying up during the summer. Such dramatic changes have caused the organisms that live there to adapt in remarkable ways. One example is the spotted salamander, which comes out during the first few rainy nights of early spring when the temperatures are above freezing. While venturing out on such a night, you will want to take a flashlight and wear warm, dry clothing to search for these cryptic animals. In the glare of oncoming headlights you may also see wood frogs and spring peepers driven by the rhythms of their mating instinct. At times the roadbed seems to have come alive with jumping and crawling amphibians. Male spotted salamanders arrive first and, when the females come, deposit sperm on submerged plants. The adult female takes the sperm into her cloaca, where the eggs are fertilized; then she lays the eggs in the vernal pool, where there are fewer predators to eat the eggs and young than would be present in permanent bodies of water. Their work completed, the salamanders return to a life in the dark world of the soil and leaf litter. [8]

SPRING PEEPING

Heard any signs of spring lately? Usually, I hear the first sound of peepers in my pond about the second week of April. In spite of fickle April's wintry interludes, spring does march on.

Those little frogs began to make their chilly way in March from their winter shelters under forest logs and leaf litter to ponds which may still have some ice islands floating in their centers. Spring peepers are known best by their lusty songs, with which the males fill the spring nights (and often days) in their efforts to bring their later-sleeping lady friends to the breeding ponds. Many people have never *seen* a peeper, and it's no wonder. The loudly vocal creatures are remarkably tiny—no bigger than a human thumb—and like many frogs, they are masters of camouflage. Virtually the only way to spot one is to slowly home in on its loud peeping and catch a hint of movement as the frog's inflated vocal sac pulsates with song. Even that can be maddeningly difficult, since the little singers have a habit of becoming instantly silent when an intruder approaches. It has always amazed me how quickly and completely the whole jangling chorus can stop, even when I am still several yards from the nearest peep. It may be that the frogs don't *see* me coming but rather *feel* the vibrations of my footsteps through the ground. Research with some Puerto Rican frogs has shown that they both sense and send "seismic signals" in addition to vocal ones.

But if you're lucky, you'll spot the peeper, clinging with his oversized, sticky round toe pads to a stem or twig above the water. As he peeps, you'll see his entire throat bulging out around his mouth, almost as though he's blown a great bubble with chewing gum. Actually, the bubble is formed by the thin skin of the frog's throat and acts as a resonating chamber to amplify his call. It does a good job, too: these one-inch singers can be heard half a mile away, and an entire chorus of tiny peepers, to someone in the midst of it, can be earsplitting. [9]

WOOD FROGS

With the warming rays of the welcome sunshine, snow and ice have given way to mud and puddles everywhere, and the brooks are roaring with springtime energy. Pussy willows are in flower, redwings have been singing for several weeks, and any evening now I expect to hear the calling of frogs from my pond, announcing their re-emergence from winter hibernation and advertising amorous interests.

Wood frogs are among the earliest to appear, and are sometimes heard before the ice is gone from the ponds. The wood frog's call is a low "quacking," somewhat like the quack of a duck. It is the males who call, offering a courtship invitation—and a directional signal—to the female frogs, who follow the quackers from woods to pond. The calls are produced when the frog takes a mouthful of air, seals off his mouth and nostrils, and "pumps" the air across his vocal cords, from mouth to lungs and back. Two pockets, or vocal sacs, in the mouth lining act as resonating chambers for the sound.

Wood frogs are so named because they live in or near wet woods. Adults may be found considerable distances from water and, unlike many other amphibians, they do not hibernate in the soft mud at the bottom of ponds, but rather under stones, logs, or stumps in the forest. With the return of warm weather, however, the frogs migrate to ponds or temporary breeding pools to mate and lay their eggs.

It is essential that the eggs remain wet if they are to develop and hatch. The jellylike mass, usually attached to submerged vegetation at the edge of the pond or pool, is subject to the extremes of early-spring temperatures. The eggs are vulnerable not only to freezing and drying, but to fungus diseases and predation as well. Fortunately the wood frog lays many eggs (up to 2,500 per mass) and they develop very quickly. It takes between six and ten days, depending on temperature and moisture, for the eggs to hatch. The tiny creatures that wriggle out of the jelly eggs are little more than heads and tails. But these tadpoles, or "pollywogs," as the larvae of frogs and toads are called, will undergo an astounding series of developmental changes. From strictly aquatic, plant-eating, fishlike swimmers they will become wood frogs in a matter of two or three months. They first develop internal gills, then lungs. Little buds on the tadpole's sides become back legs, and the gills are lost as the front legs push their way out, closing the former gill openings. At this point the tadpole tail disappears—it is actually reabsorbed as nourishment by the body of the "froglet."

Now a land-dwelling, air-breathing, wholly carnivorous hopper, it is hard to believe that this animal and the tiny tadpole that was swimming in the pond a short time ago are one and the same creature. [10]

TURTLES

Once turtles reach adulthood, their longevity provides them an opportunity to reproduce for many years. Many freshwater species, such as spotted and painted turtles, live into their twenties. Individual turtles may live even longer; one spotted turtle lived for forty-two years in captivity. Some species live thirty to fifty years or more in the wild. Determining the age of a snapping turtle after twenty has proven difficult, but doubtless they exceed this considerably. The eastern box turtle, which has been known to surpass the age of one hundred twenty, is the longest-lived vertebrate in North America.

March, April, and May are the primary months for turtle courtship and mating. The individual habits of different species and factors of climate and latitude can extend the overall period in North America from February to November. Generally turtles do not mate until water temperatures have moderated and feeding has begun. Northern species at the Canadian border are still locked in hibernation under great mantles of ice when males and females of southern relatives are pairing off along the Gulf Coast.

More tolerant of cold water than most species, spotted turtles have been known to breed in early March, with water temperatures at 46°F and an air temperature of 54° F. Courtship and mating in snapping turtles begin in April, and although most breeding activity has been completed by the end of May, pairs have been observed mating into November. Female snappers are among those species that can store live sperm and produce fertile eggs in May or June as a result of a mating late in the previous season.

Turtles can take life-sustaining processes to limits that are difficult to imagine. To their extraordinary abilities to go without food (and in the case of tortoises, water) for months or to survive half a year without access to air can be added remarkable reproductive resiliency: females of some species can store viable sperm for periods of years. An eastern box turtle or diamondback terrapin, for example, prevented by environmental or other interventions from contact with a male, can lay fertile eggs in a nesting season four years after her last mating. Turtles are unusual in that production of sperm by the male and eggs by the female are not synchronized. Eggs begin to develop in the female in the fall, when sperm production in the male is coming to an end. The sperm cells with which the male fertilizes a female in the spring have been produced the preceding year and kept viable within his body through the winter's hibernation.   [11]

OUTWITTING BUSY BEAVERS

For decades, game wardens, wildlife biologists, and highway maintenance crews have spent countless hours and dollars removing beaver dams and repairing the damage they cause. This has usually meant digging out tons of intertwined sticks and mud the beavers used to plug culverts and bridges in creating the backed-up ponds in which they construct their lodges and winter food piles. Making matters worse, the ambitious rodents also have the habit of rebuilding overnight what humans have spent days removing.

Digging out or dynamiting a beaver's dam can also be counterproductive, as there are benefits deriving from beaver ponds if their high-water level can be controlled. Beaver ponds and marshes make ideal habitat for a variety of wildlife species, including waterfowl, aquatic mammals, fish, moose, and deer. They also provide a source of water for use by rural fire departments.

One man who spent years of his professional career working on beaver problems is James Dorso of Gardiner, Maine, a wildlife technician for the Department of Inland Fisheries and Wildlife until his recent retirement. In his twenty years with the department, Dorso not only dug out numerous beaver dams, but also routinely live-trapped beavers in nuisance areas and released them in places where they would not contribute to problems, and where their ponds could create new wildlife habitat.

After years of experimenting with water-level control techniques, Dorso finally came up with a technique which is both labor- and cost-effective—and works.

It is, essentially, a framework of metal fence posts (supplied by the Department of Transportation) on which are laid an array of five or six four-inch perforated plastic pipes. The downstream ends are anchored low in the beaver dam structure, and the upstream ends of the twenty-foot pipes are held at the desired level over deep water. The pipes are laid side by side, several feet apart, and a protective "cage" is fashioned over the ends to keep away floating debris.

Dorso says the siphon pipes continue flowing water through the dams no matter how high the beavers build them, thus maintaining optimum water levels for wildlife while protecting roads and timberlands.

Dorso constructed several of his "beaver foilers" on flowages this year. "So far, the beavers have been stymied," he says. [12]

THE OTTER—
NORTHWOODS PLAYBOY

The otter, a large weasel-like mammal known scientifically as *Lutra canadensis*, is perhaps the most aquatic of our "land" mammals. With a sleek, powerful body and webbed feet both fore and aft, he is a skillful swimmer able to catch a fish with relative ease. An otter can stay submerged for up to four minutes before resurfacing for air. His long, tapered tail is used for steering. Male otters are larger than females, and adult males weigh an average of twenty pounds, with large specimens sometimes topping thirty pounds. The otter is well known for its beautiful dark brown fur, which consists of a dense layer of underfur protected by shiny dark brown guard hairs.

Otters are extremely playful and like to roll around on the ice and climb over each other. They are famous for their "slides" in which they repeatedly slide down steep river and stream banks into the water. When there is a good cover of snow on the ground, otters invariably slide down even the slightest incline. They do this by folding back their front feet and pushing off with their rear feet much like a toboggan load of kids. With their short, stubby legs, otters would apparently rather slide over the snow than walk. Who could blame them for that?

More often than not, otters travel in groups, usually of

two to five animals and sometimes more. They are very sociable and seem to enjoy the company of others of their kind.

Otters are occasionally known to prey on muskrats and young beaver that have wandered too far from the protection of the lodge and their parents. Clams, frogs, and crayfish are also part of the otter's menu.

Otters are occasionally controversial in the North Country because at times they prey on trout; there are undoubtedly hatchery operators and small trout pond operators who could attest to that. In a more natural setting, however, trout do not appear to be high on the otter's list of preferred food items. Through analysis of stomach contents and droppings, biologists have found that rough fish like suckers, chubs, and bullheads occur most frequently in the otter's diet.

Otters have a regular circuit of waterways which they follow in making their way about their home range, which is said to be approximately sixty square miles. They have a series of dens along their route that are used as stopover points. Beaver lodges that were abandoned when the food supply became exhausted are favorite resting places for traveling otter. They also do not hesitate to travel cross-country, but if one follows such an otter track, it invariably comes to another watershed.

The winter coating of ice on streams and ponds is not an impediment to the otter. If no air spaces are to be found between the ice and water, the otter will search out a crack along the shore where it can come up for air. Some holes in the ice are kept open all winter by the otter.   [13]

SOAKING IT ALL IN

It's been said that there are few more successful animals on earth than sponges. They have existed largely unchanged for at least 500 million years, I found while sailing among the Greek islands.

I find that throughout the world there are more than five thousand known species, and, undoubtedly, many others exist that we don't know about. They range from polar to tropical seas, from just below the surface to depths of several thousand feet. Some are fragile, tiny animals no larger than a grain of rice. Others branch out for many feet and weigh as much as one hundred pounds. A few live in freshwater lakes and rivers. All of them are among the earth's most primitive multicellular animals: they have no hearts, brains, lungs, or nervous systems.

Despite the variety of shapes and sizes, all sponges have the same basic structure. Their bodies are "punctured" by dozens of pores that suck seawater into an internal canal system. Inside, the animals have special cells that sift out food particles. Those substances that the creatures cannot digest are ejected. Sponges reproduce by releasing clouds of sperm or eggs into the water, and parts of these clouds are eventually consumed by other sponges of the same species.

Because they taste so bad, sponges have few natural enemies. However, they do offer shelter to a variety of sea creatures. The Venus Flower Basket, for instance, a delicate white sponge found in the Pacific Ocean, frequently houses a pair of young male and female shrimp. As the shrimp grow, they become trapped inside the sponge and sometimes spend their entire lives there, feeding on plankton sucked into their host. Hermit crabs, on the other hand, occasionally attach small pieces of sponge onto their shells. As the sponge grows, it provides an additional protective covering for the crab against predators.

Some sponges are seemingly aggressive; they will bore holes in coral or shellfish in search of food, and damage the creatures in the process. Most sponges, though, lead very passive lives. While several species live for decades or longer, they may spend their entire lives in one place on the ocean floor.   [14]

FISH SOUNDS

Dolphins, of course, are not the only sound-producing creatures in the sea. At least 83 of the 117 known species of marine mammals (including sea otters, pinnipeds, cetaceans, and sirenians) have been tested, and all of them make noises of some kind—yelps, whistles, screams, raucous blares, squawks, rasps, mews, trills, whines, or moans.

The lower orders are not silent either. Spiny lobsters, for instance, produce loud raucous bursts of sound by rubbing the ridged bases of their antennae against toothed surfaces on their carapace. The American lobster sometimes growls. Some species of fiddler crab emit a low humming associated with both threatening and sexual behavior, and one kind of ghost crab can make a hissing sound. Black mussels produce a crackling by breaking the elastic byssal threads anchoring them to surfaces.

Fish sounds vary greatly in frequency range, pattern, principal components, amplitude, duration, and repetition rate. Some species display a "vocabulary" of more than one signal, particularly in different geographical areas or during breeding season.

In many fishes, sound is produced by the contraction of muscles associated with the swim bladder. Other species sound off by grinding their teeth or by rubbing against each other patches of mosaiclike teeth located far back in the pharynx.

The majority of fishes with swim bladders produce an "escape" sound, usually a quick "knock" with occasional repetitions. The sound is produced when the fish is startled, or just before it twists suddenly during rapid evasive acceleration.

Here are sounds made by some familiar fishes:

- *Scup*—low guttural grunts, knocks, and stridulatory rasping during competitive feeding; scrunching and grinding noises during feeding

- *Weakfish*—only males make noise; deep thumps like drumbeats and bursts of higher-pitched croaking
- *Bluefish*—weak clicks and thumps
- *Striped Bass*—low "unk" with tom-tom quality, single or in bursts of three when alarmed; thumps, clicks, scrapes under duress
- *Atlantic Cod*—low growls and thumps
- *American Eel*—long-continued, low clucking which resembles the "putt-putt" of an outboard motor; weak clicks and squeals in air on capture; low knocks and underwater clicking during nighttime activity of groups

In *Historia Animalium*, Aristotle noted that, out of water, three species make a piping sound, another species growls, yet another grunts, and one makes an out-of-water sound resembling that of the cuckoo.

Cuckoo? Why not? If eels can cluck, why can't a fish cuckoo?!   [15]

SCHOOLING

Of the twenty thousand species of fishes, more than half swim in schools at one time or another. Although it is a form of animal social behavior that must have been familiar to ancient man, schooling, until recently, has remained a puzzling phenomenon.

Why do fish school? What leads the school? How do fish maintain position within a school?

Studies over the past few years have solved some of the unknowns about schooling or provided data upon which to base speculative but plausible explanations.

It takes at least three to form a school. When two fishes travel together, one leads and the other follows, adjusting its speed and direction to match those of the leader. The lead fish, however, makes no adjustments in response to movements of the following fish. When the two are joined by a third fish, the pattern changes. Then no fish "leads" and each adjusts its speed and heading to agree with those of the other two. In effect, the school is the leader and the members of it are the followers. The same dynamics obtain in a large school, with each fish adjusting its track in response to the movements of its nearest neighbors.

Some species spend more time schooling than others. Herring, for example, spend most of their time swimming together (and are called obligate schoolers), whereas cod and minnows congregate only some of the time (facultative schoolers).

Another common characteristic of schools is their polarization—the parallel arrangement of their members. In a group of feeding fish, individuals face in many different directions. When such fish travel together, however, they most often dispose themselves in a polarized pattern. When a school is threatened, individuals crowd closer together, frequently aligning themselves more precisely with their neighbors.

One obvious advantage of schooling is that a group can search a larger area than an individual can, so that food-finding is easier for a schooling fish. But schooling also helps fish to meet another urgent need—escaping predators.

A fish increases its risk of detection only slightly by joining a school. For taking that risk, it gains the advantage of numbers. If, for example, a barracuda takes 3 fish from a school of 1,000, the chances of being eaten are only 3 in 1,000 for each member of the school. A single fish detected by a predator obviously has a much slimmer chance of surviving.   [16]

THE ELEGANT TUNAS

The scatterbrain forces of evolution have not been kind to all sea creatures. Consider the flounder, knocked flat and cockeyed by a wild gene and forced to swim on its side; or the goosefish, all head and uglier than a paper dragon; or that giant pancake of a fish, the mola mola, which forces its bulk through the deep with pitifully small fins and a fleshy flap for a tail.

Nature did right by the tunas, however. In the same family with swordfish, marlin, and sailfish, tunas are among the elite of the world's twenty thousand fish species. Their torpedo-shaped bodies, with retractable fins set in grooves, are marvels of hydrodynamic design. Propelled by powerful tails, they streak through the water fast enough to catch almost anything that swims and faster than almost all other creatures with a tooth for tuna. They are rare among fishes, moreover, for being warm-blooded, able to maintain a body temperature higher than the temperature of surrounding water. This enables tunas to convert food rapidly to the energy they need for constant movement, respiration, and bursts of speed. It allows them to range through different oceanic environments in search of food—from the warm waters of the Gulf of Mexico, for example, to the chill depths off Nova Scotia.

Of course, tunas are not only superbly structured, they are also supremely tasty. Of thirteen tuna species, five have great commercial value: the bluefin and the albacore of temperate northern and southern latitudes; the skipjack, yellowfin, and bigeye, widely dispersed in tropical waters. Man's taste for tuna is an ancient one. On the wall of a cave above the Bay of Biscay is the likeness of a bluefin tuna drawn by an Ice Age artist whose people fished for it with primitive methods. Incas and other ancient peoples along the Pacific coast of South America also harvested tunas, as did the Greeks and Romans in their quarter of the world. The Phoenicians were gathering bluefin in traps three thousand years ago in the Strait of Gibraltar.

Today seventy nations have tuna fisheries which use methods ranging from pole-and-line fishing to large, purse-seine liners. Modern technology—nylon nets and power blocks, helicopters, and massive freezing lockers—have made present-day tuna harvesting both easier and more devastating to stocks than it was before the turn of the century. Yet in most cases finding the tunas is almost as much a matter of shrewd guesswork and luck as it was in pre-modern times, for we do not know a great deal more about these quicksilver sea-roamers than our ancestors did.   [17]

LANDLOCKED SALMON

The landlocked salmon (*Salmo salar*) is one of the most prized cold-water fishes in the Northeast. Modern ichthyologists agree that there are few physical differences between the landlocked salmon, which spends its entire life in fresh water, and the sea-run Atlantic salmon, which spends most of its adult life in salt water.

Many theories have been advanced on the origin and development of the landlocked strain. The most widely accepted opinion is that the preference for fresh water was a physiological change that took place through evolution. In nearly all the waters where the original landlocked salmon were found, access to the sea was available. Landlocked salmon are generally found in deep, cold, well-oxygenated lakes where there is little competition from other species.

Salmon spawn in October and November, using inlet or outlet streams. Lake outlets that have high dams prevent the return of adult and young salmon to the lake, and these should be provided with fishways. Female salmon generally reach sexual maturity during their fourth or fifth year. Males develop faster, maturing by their third and fourth year and quite commonly even as one- and two-year-olds.

Areas preferred by salmon for spawning consist of gravel

riffle areas in swift-moving water. The female builds a nest or egg pit in coarse gravel or stones ranging in size up to one and a half inches in diameter. The eggs are deposited in the gravel at a depth of four to twelve inches. Several egg pits may be dug by a female to contain her eggs, and these areas are collectively known as a "redd." A single female usually produces 600–700 eggs per pound of body weight. The incubation period averages 150 days, although it varies according to water temperature. After hatching, sac fry remain in the gravel until their yolk sac is absorbed, usually early in June. The young then move into riffle and boulder or rubble areas, feeding mostly on insects.

One to three years are spent in the stream before the young salmon migrate to the lake. Once in the lake, their growth rate increases rapidly, and their diet changes mainly to fish. Growth rates may vary from lake to lake, but best results are obtained where smelt are abundant.   [18]

FIERCE FISH

The ocean is full of such ferocious beasts as killer whales, barracudas, and great white sharks. But the most vicious freshwater animal in the world is a small, silvery fish—the bloodthirsty piranha.

Piranhas, which inhabit the lakes and rivers of South America, attack their prey without warning and can quickly turn animals of any size into skeletons. Hundreds of the fish attack at once with lightning speed, their razor-sharp teeth slicing off chunks of meat as they dart around their victims, chopping and swallowing. Sometimes piranhas get so excited that they start a feeding frenzy, gobbling up anything they can grab—even each other.

The piranha is well equipped to be a deadly predator. Its sharp, pointed teeth snap together like a steel trap, and strong, bony plates on its head protect the fish as it smacks against the body of its prey. In addition, these savage fish have very keen senses of smell. If a wounded animal is in the water, piranhas will race toward it from all directions in an instant. Most piranhas, however, eat other fish.

Scientists have discovered a strange thing about piranhas: they may be fierce killers in one river, but as harmless as goldfish in another. And while some varieties of the fish are so dangerous that a person trailing his hand in a South American river may lose a finger, not all piranhas are hazardous to people.

South American Indians living along the Amazon catch and eat the tasty fish and can tell the different varieties apart by the various smears of color along the fishes' bellies. They call one kind of piranha the "kitchen boy." When villagers leave their dishes in the river, these fish nibble at the leftovers. As soon as an Indian catches a piranha, he grabs the fish behind the gills, raises it to his mouth, and bites the fish behind the head, quickly breaking its spine. After a meal of roasted piranhas, the Indians use the fishes' jaws as knives or scissors.

Many states, fearing that piranha owners might set their pets free in lakes and rivers, have passed laws making it illegal to sell the ferocious fish. The best place to view piranhas is in an aquarium. Just don't wiggle your fingers in the tank.   [19]

GOLDFISH

A goldfish in a bowl offers a convenient introduction to fish—the way they move, breathe, and live in perfect accord with their aquatic medium. If you watch a goldfish swim, for instance, you will notice that fish have the attributes of both snakes and jets. The goldfish wiggles its lithe body as if it were an extremely active snake, propelling its streamlined shape forward as if it were an extremely maneuverable jet.

The large tail fin provides the main thrust, while the other fins serve to turn, lift, stabilize, and stop the fish. The pair of fins near the head—those that are the counterparts of human arms—provide direction. The pair on its belly elevate the fish, but they also roll it slightly from side to side. This tendency to roll is corrected by the fin on the fish's back and the one just in front of its tail. If the fish wants to stop, it uses its "arm fins" as brakes, and if it wants to rest in a stationary position, it flaps these same fins rhythmically to counteract the rhythmic spurts of water coming from the openings on the sides of its head.

As you watch a goldfish hovering gracefully in one spot, you will see that it seems to be mouthing the words, "yup, yup, yup." This repeated opening and closing of the mouth is the way a fish breathes. It draws in some water, which passes over the gills—delicate, thin-skinned fringes filled with small blood vessels. Oxygen from the water can move right through the thin skin directly into the blood, which transports it to the rest of the body. Wastes, such as carbon dioxide, can move the other way, from the blood through the thin skin into the water, then out through the gill slits, the openings that show quite visibly on the sides of the goldfish's head.

Everything else about the goldfish is adapted to its watery existence too. Overlapping scales lubricated with mucus enable it to move efficiently through water. Its large, lidless eyes, which researchers speculate are nearsighted, enable it to see far enough to detect food or danger, and its nose, which is totally unrelated to its breathing apparatus, enables it to extract scents from the water that washes through its small nostrils.

Goldfish have become popular indoor pets because they adapt more readily to the vagaries of human care than do most other fish. Although a goldfish appreciates a filtered and aerated aquarium, it can survive in a bowl of tap water as long as the water is fresh and not too soft, hard, or chlorinated. It tolerates room-temperature water—even if it prefers water that's cooler than most rooms—and will eat almost anything you feed it. A goldfish is an omnivore—an eater of a wide variety of plant and animal foods—who needs regular feedings of commercial goldfish food to balance its diet but will also eat bread crumbs and cereal (which are not especially good for it) and small pieces of earthworm, insects, and bits of plants. [20]

KILLER WHALE

Largest member of the dolphin family Dolphinidae, the killer whale belongs to the order Cetacea and travels in all oceans of the world. Highly social, it spends its life as part of tightly knit family groups called herds or pods, which usually number five to twenty members. Each pod appears temporally stable, composed of the same individuals—newborns typically remain with the pod throughout life.

The killer whale is well equipped for life in the sea. Its streamlined, hydrodynamic body is covered with very smooth skin, similar in touch to wet rubber. A triangular-shaped dorsal fin (growing five feet tall on males) is located near the middle of the back. The pectorals contain the same bones as human hands, wrists, and arms.

The killer whale often pops straight up out of the water to visually survey its topside surroundings. In areas where underwater visibility is limited, the killer whale relies on a sonarlike ability to examine its environment and finds prey by echolocation. It emits a series of sounds that travel from a portion of its head called the melon and reflect off objects around it, producing echoes it can hear and interpret. The whale receives almost instantaneous information about prey, including its size, shape, direction, and distance.

The killer whale is primarily jet black on top and brilliant white below. This countershading is an effective camouflage for both hunter and hunted. Observed from below, white undersides blend with light skies—viewed from above, dark bodies blend with the blackness of the deep ocean.

Having no natural predators, the killer whale dominates

the top of the oceanic food chain. Clocked at speeds up to about thirty knots, it can chase down any prey in the sea. The killer whale feeds on a variety of marine life, including fish, birds, squid, seals, and sea lions. Amazingly, the killer whale may attack and devour other whales, including the great hundred-foot-long blue whale. Consuming an estimated three hundred pounds of food every day, the male killer whale can grow up to thirty feet long and weigh over eight tons.   [21]

SQUID

The squid belongs to the phylum Mollusca, which includes the more familiar shellfish—clams, oysters, scallops, mussels, snails. The squid shell has been modified over 400 to 500 million years of adaptation into an internal remnant, a rigid longitudinal rod.

Squids of some kind are found in almost all waters, at depths ranging from inches to more than 15,000 feet. They move about in search of food, not by swimming or by crawling along the bottom with their arms as the octopus sometimes does, but by jet propulsion. The squid body, or mantle, is a soft, muscular cylinder projecting straight back from the head and arms. On either side of the mantle is a flat oval tapering off into a pair of stubby, winglike fins. Into the mantle cavity the squid draws water, which it then ejects through a multidirectional spout, or funnel, thrusting itself in the desired direction. When alarmed, the squid is capable of a sudden, powerful contraction of the mantle that rockets the creature backwards out of harm's way.

Watching a squid hover in place or glide casually through the water might cause one to wonder how the creature ever got a reputation as a predator. But for complacent fish, the squid has a surprise. Amid its eight arms are two additional tentacles that resemble the arms until they go into action. Then they prove to be remarkably fast and elastic, lashing out from among the squid's arms to grapple its startled prey with clusters of suckers or hoods. The tentacles then retract, drawing the unfortunate fish within reach of the grasping arms, which wrestle the prey to the mouth and hold it there as the squid tears at it with its parrotlike beak.

Squid sex is indeed a climactic experience, occurring but once in a lifetime. Its mechanisms, however, are wonderfully varied, as any intimacy involving 16 arms and 4 tentacles would have to be. In the midst of a complex, ritualized dance of passion, the male uses a specially adapted arm to withdraw from his mantle cavity a packet of sperm which he then deposits in the female's receptacle—a pocket under the eye on one species, a cavity on the neck in another.

The complicated intensity of the experience would seem to be nature's way of compensating for a raw deal: immediately after the strings of fertilized eggs are attached in moplike bundles to a rock or left in the form of sausages to float in the open sea, both male and female usually die.   [22]

THE LOBSTER

When wolves roamed New England and virgin forests were dark with huge trees, the American lobster, *Homarus americanus*, abounded in rock pools and shallows along the coastline. Losing only a few of their number to Indians and settlers, these crustaceans grew to patriarchal sizes, commonly weighing twenty-five pounds. Farther north in its range, the American lobster was so plentiful that early farmers in New Brunswick used it for fertilizer. The days of plenty did not last long, however.

The American lobster fishery began in earnest in the late 1700s, mainly off the coast of Massachusetts. As early as 1812 stocks had declined sharply enough to move worried Provincetown fishermen to regulate lobster harvesting. But years of taking lobsters of all sizes and even their eggs (regarded as a "capital article of food") had all but wiped out the stocks of *Homarus americanus*.

The American lobster ranges from Labrador to the Carolinas (although found in greatest numbers from Cape Cod north to Nova Scotia and New Brunswick), and populations were available to fishermen elsewhere.

Pollution and cooling trends in the Atlantic are partly responsible for the decline in lobsters, but the main cause seems to be overexploitation brought about by the increased efficiency of harvesting methods, offshore dragging by foreign fishing vessels, inadequate size-requirements for "keepers," and poor enforcement of catch regulations.

Observation of *Homarus americanus* in the wild is not yet adequate to support hard generalizations about its mating behavior, but it appears that when the female is due to molt, she approaches a male's den and releases a sexual attractant. The male responds by allowing the female to enter his den, where she soon molts. Once this happens, mating takes place. Mating completed, the couple live together for a week to 10 days, possibly to afford the female protection while her new shell is hardening.

At this point the fertilization process has just begun. The sperm remains in the female's seminal receptacle for nine to twelve months while the eggs prepare for fertilization. At the proper moment, the female extrudes up to sixty thousand eggs and cements them to the hairs of the swimmerets under her tail. As the eggs are extruded, sperm is released to fertilize them. The eggs will remain glued to the female for another nine months before they hatch from under her tail and float away, looking more like shrimp or mosquito larvae than lobsters.   [23]

LOBSTERING IN WINTER

You must have read accounts of fishing or sailing on the North Atlantic in the teeth of a gale. To that impression, add winter temperatures so low that salt water freezes to the ship's superstructure and lines, endangering the welfare of all aboard with the added, top-heavy weight. Then there's the swirling, blinding phenomenon of "sea smoke," winter vapor of the water's surface. What tales! What folks!

There is a very special, strictly six-month, protected lobster season on Monhegan Island, Maine. Monhegan is actually a submerged mountain twelve miles out to sea; her waters are very deep and, hence, moderated in water temperature a bit.

This special closed lobster season runs from January 1 to June 25. As a rule, New Year's Eve is a quiet night here— the exception is when even one person's poor health or the unworkable condition of a boat or a horrid weather forecast postpones the hoped-for Trap Setting Day of January first.

The small fleet of a dozen or so boats (the largest of which is forty-five feet long) waits until everyone is ready, able, and all boats loaded with stacked lobster traps before "they're off," usually with the blast of a horn.

In this winter community of about seventy, the dock area is abuzz with life as most everyone is called into action to help drive the trucks laden with traps, bait, and buoys, to actually loading the boats that take turns at the dock's edge. "Dinner pails" have been packed, baked goods and hot beverages are available, and a good spirit is in evidence.

Months of work preceded this event, involving the boats, the gear of traps, lines, and buoys—each buoy painted with the colors and mark of its owner. It is all hard work, especially in winter conditions. The generally present windchill increases the perceived sense that it *is* unbelievable. But those who choose to lobster on Monhegan wouldn't have it any other way—there seems to be an indescribable love or comfort that is born in independence and having no interface with your Maker. It is you, the sea, and the sky, the elements and the creatures of the air and the deep.   [24]

PLACES

REINDEER IN LAPLAND

Christmas is close at hand. Strains of "Rudolph" drift through December's air bringing memories of the real reindeer we saw in Lapland.

One fall we were steaming up the coast of Norway on a mail boat that brought supplies and passengers to isolated hamlets approachable only by sea. On a cold windy afternoon as the sun was setting at 2:30, for we were near the North Cape and far above the Arctic Circle, the captain invited us, the only American passengers, to the bridge for a visit. The view up there of the North Sea was impressive as we watched the ship bury its bow in the cold gray seas. There were kittiwakes and Arctic skuas to look at, but what held our attention were the many grayish-brown beasts grazing on withered grass atop a seaside cliff. They were reindeer, said the captain, and were migrating, away from their exposed summer feeding grounds to more protected inland areas to spend the winter months.

After we had left the ship and were crossing the boundary river between Norway and Finland on a raft, we met our first Lapp, the traditional herders of reindeer. The only other person crossing on the raft with us, he was dressed in Lappish costume and carried over his shoulder the skin of a reindeer which must have only recently been parted from its owner, for it oozed blood on our boots.

We soon became accustomed to seeing herds of the wandering animals, pawing through the newly fallen snow with their big shovel-like hooves, followed by their herders. They are all-purpose beasts, supplying milk, meat, and skins for the Lapps. The Lapps eat well, for nothing is more elegant than smoked reindeer tongue. Alas, however, the ubiquitous snowmobile has taken the place of the reindeer in pulling sleds.

The last glimpse we saw of reindeer was in a snow storm, late at night, in the headlights of the bus we were taking to southern Finland. Transfixed, the big stag stood stolidly in the middle of the narrow icy road. The driver wrestled with the wheel, missed the reindeer, but we ended up in a ditch. "All out," said the driver, "and push the bus." At least I guess that's what was said, for it was all in Finnish. I got out with the rest of the men and we collectively put our shoulders to the bus, grunting, "One, two, three" in Finnish, Lappish, and English. I had to smile. There I was, an American, pushing with huge Finnish loggers from the forests and small Lapp reindeer herders in their strange dress and reindeer boots, none of us understanding each others' speech, in the middle of the night in a ditch. It was such a funny adventure that I've always been grateful to that big old reindeer.   [1]

ICELAND

As the plane swung around to land, steam was rising from hot springs surrounding the airport; in the distance glaciers shone in the early-morning sun. Iceland was living up to its reputation as the land of fire and ice. Ever since we had briefly stopped over in Reykjavik a number of years ago, we had vowed to return to this strange island.

Iceland, in spite of its name, is not a cold country. Because the Gulf Stream washes by its shores, Iceland's summers are cool and its winters mild. It is, however, the most volcanic place on earth. Its fiery nature is due to its being, in geologic terms, a brand-new island, the meeting place where the North American and European tectonic plates grind together to spew out fire and lava. The austerely beautiful Icelandic landscape is dotted with groaning, gurgling mud holes, weird lava formations, roaring waterfalls and geysers, all topped with lofty black and white mountains. An Icelandic friend reminded us that America's astronauts practiced walking on Iceland's lava fields before they flew off to the moon.

But all is not black and white; between the mountains are fields of the brightest green where sheep and fat Icelandic ponies, descendants of those brought over by the Norsemen, graze by blue lakes. Greenhouses, heated by thermal hot water, supply the islanders with fresh vegetables, tomatoes, and even bananas!

For years, until the Second World War, Iceland, tucked away in the upper reaches of the North Atlantic, was virtually ignored by the rest of the world and was happy to have it so. Then, afraid that Iceland would fall into Hitler's hands, Britain and the United States installed troops on the island. Willy-nilly, Iceland was dragged into the twentieth century.

The people, proud of the fact that theirs is the oldest democracy in the world, over a thousand years old, have coped successfully with the change, enjoying one of the highest standards of living in the world with no illiteracy and very little crime.

It was no use trying to sleep in Iceland's midnight sun. We wandered about the little port of Seydisfydjour, from whence we were to take the overnight ferry to the Faeroe Islands the next day and counted forty waterfalls roaring down from the glacier-clad mountains on either side of the narrow fjord. Soon it was 1 A.M. and time for bed. We pulled the curtains, shutting out the sun, and went to sleep dreaming of geysers, glaciers, and waterfalls. [2]

HARBOR WATCHING

There are few better noncompetitive sports than harbor watching. For the landlubber there's the fascination of watching seamen going about their arcane tasks; for those with wanderlust in their hearts there's the temptation to throw everything aside and join a seagoing vessel to some distant port. But the harbor has to have the right amenities to qualify to be a proper harbor-watching spot. It mustn't be one catering only to yachts; there have to be honest, hardworking craft that have real reason to be at sea.

One of the best harbors Jane and I have found is on the faraway Faeroe Islands. We could hardly believe our luck when we sailed into Torshavn from Iceland one June day. Torshavn is the capital of the Faeroe Islands in the North Atlantic, midway between Iceland and Norway. The capacious harbor was jammed with every imaginable size and shape of boat, working boats, oceangoing ferries, rowing boats, gunboats, cargo ships, huge deep-sea fishing boats. Most elegant were the wooden in-shore fishing craft built just as the Vikings had built theirs a thousand years ago; the only difference was an engine.

Every morning we'd stroll down to the docks to oversee the unloading of a catch of fish, watch the Royal Danish Navy scrub its decks, or see who had arrived on the ferry from the Shetland Islands. Sometimes we ate our picnic lunch with our backs against Torshavn's medieval buildings perched on a spit of land jutting into the harbor. The grassy ramparts of the fort was another favorite picnic spot. British troops had been stationed in the fort during the Second World War, for the Faeroes were a vital link in the Battle of the Atlantic.

An archipelago of eighteen islands, the Faeroes are an independent entity, but Denmark guards its rich fishing waters and is responsible for prosecuting crime. Not many to worry about; when we were there Torshavn was buzzing with talk of the first bank robbery it had had in fifteen years.

With its population of 15,000, Torshavn, the smallest capital in the world, is a wonderful mix of the ancient and the modern; we were delighted to see that the roofs of the new college buildings were covered with grass!

We liked Torshavn so much we even picked out a house to buy; no matter it wasn't for sale, we just liked to dream of sitting in a cozy house with a grass-topped roof and looking out at that busy harbor. [3]

THE RIGHT ANIMAL
IN THE RIGHT PLACE

We've always been interested in observing how well animals, domesticated or wild, suit the place in which they live. The Herdwick sheep we saw in England's Lake District appear to be extraordinarily well adapted to survive in the area's rugged, treeless, high fells. Where they came from nobody is sure, but they probably came over from Norway with the Vikings who settled there a thousand years ago.

Herdwicks look misleadingly meek with white faces and legs and a dense fleece of coarse gray wool. But it is this coat, full of wiry fiber called "kemp," that effectively keeps out the rain that on the fells may amount to two hundred inches a year. Winters, too, can be cruel on the mountainsides. The tough Herdwicks have been known to survive up to three weeks though completely buried in snowdrifts.

Another characteristic that makes these animals suitable for their inhospitable, barren habitat is their unique awareness of the boundaries of their native pasture. The fell pastures are unfenced, and animals wander at will over immense open areas. In the Lake District this awareness is called the "heaf," used in the same way as when we speak of "our native heath." This makes life a trifle easier for the shepherds, knowing that their flocks will stay on their own pasture and not wander into the next county.

Once the shepherd brings his sheep down to the home farm for shearing he had better have stout, high enclosures, for six-foot walls are needed to keep the strong-legged Herdwicks in the farmyard.

At the turn of the century the continued existence of this remarkable animal was in doubt, for their wool was considered too coarse for fine cloth. It was Beatrix Potter, with royalties from her children's books, who saved the sheep from extinction. In middle age she bought her beloved Hill Top Farm in the Lake District, becoming known in time as one of the foremost breeders of Herdwick sheep. For years her sturdy tweed-clad figure was a familiar one at sheep sales, urging farmers to upgrade their flocks, with the result that today the sturdy Herdwicks and their lambs are flourishing on the high fells.   [4]

THE ISLE OF JERSEY

Back when we milked cows, the name "Jersey" meant not the state or a sweater but the island in the English Channel. We raised Jersey cattle, whose forebears originated on the island. When we were in England a number of years ago, naturally we couldn't pass up an opportunity to visit the Isle of Jersey and to see the fabled Jersey cow in her own environment.

Jersey and Guernsey are part of the Channel Islands. Victor Hugo said that they were pieces of France that had been thrown into the Channel. Near the French coast, the islands are separated from the mainland by ferocious tides and treacherous sands. In the Middle Ages the islands were part of the Duchy of Normandy when the Duchy belonged to the King of England. Today they still pay allegiance to the English Crown but not to the government of England. Guernsey and Jersey have their own constitutions, make their own laws, raise their own taxes, and because the taxes are much lower than in England, the islands have become tax havens for wealthy British.

Although they have been under the influence of the English Crown for centuries, the islanders still retain much of their Norman French background. Land, for instance, is measured according to the ancient Norman system of "vergees" rather than in acres, and most of the islanders speak "Jersey" as well as French and English. The seaweed they spread on their fields they call "vraic," using an ancient Norman term.

On small pastures, still green in January, Jersey cattle

were grazing. In Vermont they would be in stables, but the island's climate is so tropical that we saw potatoes being planted and palm trees flourishing. The precious cows are led out every morning after milking, not to be let loose in the pasture, but to graze at the end of a chain. And each animal is carefully covered with her own blanket. Cows are pampered on the Channel Islands! We were told that both Jersey and Guernsey jealously guard their own breed of cattle; animals from the outside are never allowed on the islands. In fact, cows of one island are never allowed to put hoof on the other.

While we admired the sleek cattle and envied the Jersey farmers their handsome stone farmhouses, we were fascinated by the island's history, and there's enough on this one small place to spend a lifetime of study. La Hougue Bie encapsulates the story of the island. La Hougue Bie is a large grave mound dating from the Old Stone Age. Perched on top of the mound is a medieval chapel dating from the twelfth century. If that wasn't enough, the Germans who occupied the island in the Second World War built an underground hospital inside the mound!   [5]

A HIGHLAND ROAD

The road climbed steadily toward the head of the empty glen, empty except for a golden eagle soaring above our heads and on the mountainside the gray forms of sheep grazing in the heather. Only the thin sound of a lamb broke the silence. In our little rental car with its unfamiliar right-hand drive we proceeded cautiously, trying to look at the magnificent Scottish highland scenery, follow the eagle, and watch anxiously for oncoming cars. In the highlands the roads are only single track, and in order to avoid collision, you must be quick to pull into the nearest layby on your side. Then there are the sheep who lie plumb in the middle of the narrow road enjoying its warmth.

We were so busy keeping an eye on all these things that we had little time to pull into a layby in order to let an approaching large object pass. Noiselessly, on huge old wheels, an ancient open touring Rolls Royce rolled toward us, its venerable shining body, which must have seen the light of day about 1911, painted British "racing green." The enormous brass lamps stared at us haughtily but the passengers bowed politely as we stared at the vision. Strapped on behind to the "boot" was a large wicker picnic basket which held, we were sure, smoked salmon, cold meat pies, cherry tarts, and cool white wine. With a sigh we turned to our little car with its meager lunch of cheese, bread, and ale, but before we climbed in we saw a sight more wonderful than the old Rolls Royce. A shepherd on the mountainside was attempting to divide a flock of sheep, even though he couldn't see them. With two border collies directed by a series of whistles, the dogs alternately crouched, bounded forward, or crept along the ground until they had sent half the sheep up the mountain and had brought the other over the ridge to their master. Although we could see the whole operation, neither the dogs nor master could see each other until the end. Then sheep, dogs, and shepherd disappeared and we were left once more in the lonely glen, empty except for the eagle soaring overhead.   [6]

LONDON WITH GRANDSONS

When an important anniversary came up not too long ago, we celebrated it by taking our three grandsons to London. I must confess we were a trifle nervous. We imagined all sorts of troubles. Would they get tired and cranky? We should have known better. We were the ones that got tired. Would they miss their accustomed American food? They drank gallons of tea, ate mountains of scones and clotted cream, and declared that fish and chips were every bit as good as McDonald's hamburgers. We worried about dragging them around sightseeing. Instead they were afraid they might miss something. An entire day at the Tower of London wasn't long enough; they said they still hadn't seen all the armor.

We must admit there were times when we wondered if we were going to get them home alive. Keeping track of three active boys on crowded city streets wasn't easy, and forgetting to watch for left-hand traffic caused at least one close call.

The wonderful thing about traveling with youngsters is that you see things through fresh eyes, their eyes. The Crown Jewels were twice as brilliant, castle walls higher, dungeons darker, and the guards at Buckingham Palace shouted louder and marched better than ever before.

We worried before we left that the trip would be wasted on boys aged from twelve to six. It was a huge success. The oldest, fascinated by English history, says he's going to be an historian. The middle grandson was proud that he could get us around on the underground system, and the youngest will never forget the lions at Trafalgar Square and the red-coated soldiers at Buckingham Palace. All three discovered that traveling is exciting and sometimes tiring, and that they have reserves of endurance they didn't know they possessed.   [7]

HADRIAN'S WALL

"The Caesar Traianus Hadrianus Augustus, son of all his divine ancestors, decided it was necessary on the advice of the Gods to fix the boundaries of the Empire . . ."

The Caesar Traianus Hadrianus mentioned in this inscription was the Emperor Hadrian who decided that something had to be done to protect his British colonies from the constant ravages of the barbarous Scots and Picts. In A.D. 122 he came to see the far-flung northern frontier of his empire and to plan a barrier that would put an end to such harassment. His wall remains one of the most impressive monuments of the ancient world, stretching for seventy-three miles across northern England.

On a cold, drizzly day in April as we walked along the wall it was a foreboding sight, wriggling like a gray snake across a rugged landscape, rising with the hills and dipping out of sight into valleys. The land is unfriendly; northward lie vast areas of the Scottish border, rough grazing land almost deserted except for thousands of sheep. As the wind shifted veils of mist, we imagined we could see the gray form of a Roman soldier on sentry duty and feel his sense of isolation, so far from his native Spain or Africa or Syria.

For the actual work, three legions, the Second, Sixth, and the Twentieth, then stationed in Britain, were split into gangs of one hundred, each group called a century or cohort. Expert construction workers as well as crack troops, it was they who worked the quarries, laid the foundations, and built the wall.

While the wall itself is intact, the mile castles and watch-towers have gone and all that remains of the elaborate military complex that lay behind the wall are foundations of the buildings of the large forts where the legions on garrison duty were stationed.

The most completely excavated fort is at present-day Housesteads, where with a little imagination and the help of a guidebook, we were able to walk into the headquarters, where the officer of the day must have sat with his plumed helmet and sword laid to one side, through barracks with handy nearby latrines complete with flush toilets. Nor were the steam-heated baths that the sophisticated legionnaires loved, missing; they must have been much appreciated in Northumberland's cold damp climate.

Around the fort we saw the evidence of all the civilian activities that grow up around a large military establishment, providing services that soldiers demand: quick food, bars, and women. Many a young lad must have come to the fort to join up, bored with life on the moorlands of Northumbria.

To us the most poignant reminder of the soldiers who

had lived at the wall two thousand years ago was the groove worn on the edge of the stone water tank where the soldiers had sharpened their swords. It was as though they had just finished and were hurrying to form into line ready to storm out through the gates to attack a party of Scots.   [8]

ST. PATRICK'S MOUNTAIN

Back in A.D. 450, St. Patrick was a very busy man bringing Christianity to Ireland, establishing churches the length and breadth of the land. Ireland wasn't new to him; he had been carried off as a youth from England by Irish raiders to be a pig-tending slave. After many years he escaped from Ireland to France, became a monk, and decided that he would spend his life converting the pig-loving raiders to Christianity.

He must have been a persuasive missionary, for in no time at all the Irish were following him joyously. Perhaps a Druid priest or two grumbled at the loss of their following, a clan chief may have protested, but in those years no martyrs were made in Ireland. No one was killed, only St. Patrick's charioteer, and that may have been an archery accident.

St. Patrick was indeed busy, but not too busy to climb mountains. Maybe the saint needed a change from his labors and to get away from his many followers. So, one day, the legend goes, he climbed beautiful Croag Patrick, a 2,500-foot mountain that rises straight from sea level on Ireland's west coast. There he spent forty days fasting and praying.

Now, every year on the last Sunday in July, about eighty thousand barefooted pilgrims climb the mountain at dawn carrying candles, but never a cut or a bruise do they get, we were told.

When we started up Croag Patrick one year in May, we were quite alone and were perhaps fainthearted pilgrims, for we wore thick socks and heavy hiking boots. And even then the loose, knife-sharp stones took their toll on the boots. The climb wasn't as easy as it had looked, and with every two steps upward we slid back one. The glorious view out over Ireland's mountains and bays disappeared as we scrambled up into the inevitable Irish clouds. But the presence of St. Patrick was still pervasive on the mist-shrouded summit, and we peered cautiously into the chasm into which he was supposed to have cast all Irish snakes. The legend must be true, for there are to this day no snakes in all of Ireland.

[9]

SUBMARINE CANYONS

There are at least fifteen canyons carved into the edge of the continental shelf along the southern edge of Georges Bank. They lie about a hundred miles offshore in a rough semicircle that begins east of Cape Cod and curves south and west toward Block Island.

The list of species in the canyons is long. It includes sponges, anemones, corals, and annelid worms; five kinds of crab; the cleaner shrimp and the panalid shrimp; squid and sea stars. The canyons support a host of fishes; among them, the Gulf Stream flounder, the witch flounder, the four-spot flounder; goose fish, hagfish, ocean pout, greeneye, pollock, skates, cusk, grenadiers; four kinds of hake, three species of eel; the black-bellied rosefish, and the tilefish. The canyons are also a prime habitat for offshore lobsters. Some 20 to 50 percent of this population migrates onshore in spring and early summer, returning in late summer and fall.

So little was known about the canyons at the turn of the century that some scientists even questioned their existence. Biologic, geologic, and hydrographic studies, especially those completed in the last ten years, have told us much about their function in the ecosystem of Georges Bank.

The most westerly of these gorges is Veatch Canyon; the most easterly is Corsair Canyon. The largest, located roughly in the middle of the range, is Oceanographer Canyon. It is twelve hundred meters deep, and continues far out into the continental rise.

Submarine canyons resemble land canyons. They are winding, V-shaped valleys branched with tributaries and fretted with gullies. Some of the walls rise sharply, stepped with exposed outcroppings of ancient rock. Others slope gently over bare rock and plateaus layered with variegated sediments. The floors are a patchwork of rippling sand dunes, areas of clay and silt, and rougher fields of stones and boulders cast down by erosion or dropped by glacial ice. It is the geologic diversity of the walls and floors, and the nutrient-rich currents sweeping over them, which make the canyon habitats in which many animals can flourish.   [10]

UP IN THE WOODS

It was a gray, gloomy Sunday afternoon in winter when the snow was too hard for cross-country skiing that Jane decided to walk up into the woods to see what was going on. She told me at teatime that she crossed the brook which could still be heard gurgling beneath the ice and followed the trail up out of the ravine to where the pines had been selectively harvested twenty-five years ago. Brush and saplings have taken the place of the old giants, making good cover and food for birds and mammals that think of our farm as theirs. The resident herd of white-tail deer had chewed off the ends of the maple saplings, for the crusted snow was too thick to permit easy pawing for grass. Because there's easy eating in that area, a flock of wild turkeys had wandered about, leaving their distinctive three-toed tracks everywhere.

Up on the hillside where the pines were numerous she saw old grouse droppings. She hoped that the grouse hadn't gotten trapped beneath the crust during the recent ice storm. Grouse often spend the night hidden in the snow. By a little stream she noticed an elaborate mouse highway system, a maze of tunnels whose roofs had melted during a thaw.

Near the top of our land a pileated woodpecker had left beneath an enormous dead tree, chips of bark as big as dollar bills. As she climbed up toward the uppermost knoll her eye was caught by a pile of freshly upturned leaves. What could it be? Coming closer, she saw that the leaves were next to a hole, quite a large hole, going into the earth several feet, big enough for a fox. But how did he manage to dig through frozen earth? Dirty paw marks show fox tracks, but not recent ones. Why did he take so much trouble and then abandon the hole?

Pondering this mystery of the woods, she started back down when the most amazing occurrence of the afternoon happened. Springing in soaring leaps, a doe dashed across the trail and, following her, leaped not one but three fawns! Deer triplets! Could it be? Were they all hers? For a doe to have three fawns is most unusual!

Jane went on down to the house thinking that a gloomy winter afternoon had turned out to have all sorts of surprises. [11]

COUNTRY OF THE HILL FOLK

In the fall we like to visit the country of the Vermont hill folk. They have long since gone, a hundred years ago or so. But their country is there still, the high ridges, little lonely valleys, their long-forgotten roads that led to isolated farms. You can't get to the hill country by car, only on foot. Autumn is the best time to go; the leaves are fallen and one can see the bare bones of the country.

On a bright blue day we shoulder our day packs and set off with a topographical map. They were frontiersmen and women who settled here; Vermont was the frontier in 1760. They cleared a space in the forest, built a shelter, and survived somehow the first cruel years. More-substantial houses followed as they prospered, fields were cleared for sheep, and miles of stone walls put up. But by the 1880s the high country was again deserted, the trees taking over once more, the houses left to tumble into their cellar holes. The roads were abandoned, or as they say here, "thrown up."

We climb the first ridge, puffing a bit as the grade steepens. On top we pause to look down on the valley with its prosperous modern farms; already we have left the twentieth century behind. Now there's no noise but the sound of our passage through the fallen leaves. Stone walls climb beside us and wander off at angles through the woods. Down in a small valley we look for signs of an old farmhouse. There's the cellar hole, trees growing out of it, a huge pile of brick that was the central chimney. By the doorstep is an old twisted lilac. We think of the woman who brought the little sapling with her from downcountry. Did she move west with the others when they left to seek better land? Did she stay behind in this beautiful spot watching the hard-won fields go back to forest?

Walking down a farm lane we cross what was the King's Highway in King George's day. It connected the seacoast with the inland towns and it is said that Lafayette himself was driven along this road when he came back to visit in 1825. Now it is filled with dry leaves. No one passes now but the white-tailed deer.

Toward the end of the day we climb a high grassy hill. Bunker Hill it's called, because it is said that on the day of the battle the boom of the cannon, 150 miles away, could be heard on the summit. More likely it was named by owner Adam Howard, who fought there that day. In the little cemetery at the foot of the hill we find his grave. "Adam Howard, Revolutionary Soldier," it says.

We walk down an abandoned road to the darkened valley. Above, the country of the hill folk is still bright in the late-afternoon sun. [12]

A COLD DAY IN YELLOWSTONE

It is cold in Yellowstone Park this winter morning; the thermometer when we went to breakfast had its pointer on −20°. But at 8:30 it is a toasty −10° and we are, if not exactly hot, at least not cold. The outdoor experts are right: "layering" does the trick.

The fresh snow squeaks under our skis as we head to the Upper Geyser Basin through a mysterious landscape. The steam from the geysers, trapped by the cold air, has created a white misty world and transformed each needle of the lodgepole pines into intricate white lace. It is a scene right out of a fairy tale, but what a strange one. Geysers rumble and belch and boil, sometimes erupting into plumes of boiling water. What we take to be the bare branches of a tree turn out to be the huge rack of antlers of an elk resting comfortably in his snowy nest. Across the trail through the mist looms a large black boulder. We stop as the boulder raises a shaggy head. It is a cow bison with three calves. We prudently retrace our steps. Eerie wails from a coyote—the Indians call them singing dogs—signal news about food. Presently we see the animal, circling us, keeping a wary eye on us but otherwise unafraid. We pass pools of hot water, vivid blues, greens, and golds in whose centers are conduits leading down into the center of the earth. A castle made of mud suddenly groans as we glide by. All this thermal activity attracts animals and birds that would otherwise find it hard to survive the winter. Heat from pools melts the snow so that the elk, mule deer, and bison have better access to grass. But still we try not to approach them closely, not for fear that they might attack but because the effort of moving away from us will take too much from their precious store of energy.

We come to a deep hole in an expanse of fresh snow. At the bottom is a drop of blood and the imprint of a coyote's mask. Could the hole have been made by the coyote we saw? It had heard a mouse moving under two feet of snow and in one pounce had dived down for his prey and in another more bloody hole had eaten the creature.

We expect to see the large mammals, but it is an extra dividend to find in the Firestone River so many species of birds. On the misty water we see a gaggle of geese, while around the bend we come on swans, Golden Eye ducks, snipe, and even that rare little bird, a water ouzel.

The afternoon sun is starting to sink behind the great wall of the crater; it is time to leave the animals and head back to the world of people.   [13]

FALL COMES EARLY
IN DENALI NATIONAL PARK

Late one August when drowsy, dusty summer was drifting toward Autumn, we packed our boots and warm clothes and flew to Alaska. We were bound for Camp Denali set in the midst of the millions of acres that make up Denali National Park. The year before, in June, on our first visit, we had marveled at the carpets of spring wildflowers that spread across the tundra. There, moose, caribou, and grizzly bears wandered with their young. "But you must see the park in fall," we were told.

When we returned the only thing that had remained the same was majestic Mt. McKinley, robed in its eternal snowy glaciers. Everything else had changed. Where before the colors had been shades of green intermingled with the flowers' pastel hues, now the tundra was ablaze with brilliant yellows, and reds, and all the bronze tints of fall. A Turkish carpet would have been pale in comparison.

The animals, their young now grown, sensing that time was growing short, were busy preparing each in their own way for the approaching winter. Caribou drifted westward through the park toward their winter feeding grounds. Once we looked up from our cabin to see three or four of them walking slowly along a ridge not one hundred yards from where we sat, browsing as they went.

Grizzly bears had shed their straw-colored summer coat and were almost black. From the road, a safe distance away, we watched a big male, hungrily gobbling soapberries, putting on fat in preparation for his long sleep that would commence in October. He was too intent on getting his fill of the berries that, incidentally, taste just like their name, to pay any attention to us.

The mating season was at hand for moose, when the male with the most impressive rack will dominate the herd. One young animal was scratching velvet from his antlers by thrusting them violently from side to side against a willow bush, shreds of velvet dangling from his bloody horns.

The beaver, too, were hard at work, getting their dwellings in shape for winter. When a big animal gnawed down an alder in seconds, we watched it swim across the pond and stick the branch into a mud bank in the middle of a pond. The mud bank would be his larder in the months ahead. Soon the beaver would be busy plastering the top of the beaver house with mud, which, when it freezes, would make a strong roof, protection from hungry wolves. A mile farther on a large wolf trotted beside the road. We wondered if the frozen mud on the beaver house would protect the inhabitants from his big, strong paws.

On the tenth of September we woke to find snow had covered the ground. It was time to leave Denali to the long winter that lay ahead. Some animals would die of starvation or fall prey to predators, but enough would survive to carry on as they have for thousands of years.   [14]

YEAR-ROUND ALASKA

Jan. 25, −17°. The sunlight wakes the sleeping hillsides. Color returns to the landscape in the rosy maroon of the birch groves and the gray-green of the aspen islands. Spruce trees lose their dark tones and seem to relax into a friendlier green. Even the temperature relaxes a little. The earth and her creatures have survived another dark and cold trial.

Jan. 28, −24°. While on my way out to feed the sled dogs, a bucket pulling low from each arm, I am almost knocked over by the flying squirrel, which glides in midair before my very eyes. *Smack!* The squirrel hits the base of the spruce with the bird feeder hanging on it and scampers up the tree, claws a-clicking. I set my buckets down and walk close to the feeder. The light from the window is enough to let me see him, and for him to see me. The squirrel nibbles on the suet and then sits erect to observe me. He seems less nervous than the red squirrels, and a little larger. After a few minutes he's had enough and pulls himself up the tree. I don't see him glide down to the other tree, but I hear him *smack* into the trunk and claw his way up to a new takeoff. What a way to fly! Squirrels that go bump in the dark are fun to run into once in a while. I pick up the two buckets and continue out to the hungry dogs, who have little patience for the squirrelly delay in their dinner.

Jan. 29, −21°. A wave of redpolls undulates through the forest, landing on the alders, dangling upside-down on the cones. They are a cheery crew, always singing. The joy of their singing reflects the return of sunnier days. Since the winter solstice we have gained about three hours between sunrise and sunset, nearly doubling the day.

Jan. 30, −22°. Most amazing morning! Today I am baking, and with the cookstove oven up to temperature, the cabin is quite toasty. I open the window over the stove to let out some heat. With a fresh muffin to sample in my hand, I step out on the front porch to cool off.

The very instant I'm out the door, a black-capped chickadee lands on my muffin! Was this an accident? Did I walk into the dee's line of flight? I freeze in my tracks. The dee flits off to the branches of the spruce directly in front of me. I stay motionless, holding the muffin in my outstretched hand up to the bird. After a minute the tiny fellow comes back to the muffin, takes another taste, and whirls away off the knoll, singing in chickadee-dee-dees as it goes.

This is the first time a chickadee has ever come to my hand. Had it been watching the camp robbers who regularly flock to a handout? I was delighted and ran up to the shop to tell John of the event. On my way back to the cabin, I noticed a black-capped on the bird feeder. I went right up to it, speaking softly, "Hello, little dee. Are you the muffin lover?" But no answer. The dee did not return to my hand. I don't even know if it was the same bird. But I am still hoping, and I'll make it a habit to taste my muffins on the front porch.   [15]

PATAGONIAN ANDES

How welcome to our eyes were the white peaks of the Andes as they rose over the horizon of the Pampa when we drove into Argentina's Fitzroy National Park. We longed for mountain streams and wooded groves after endless miles of dusty roads stretching forever before us.

The most spectacular mountain in the chain of the Patagonian Andes is Mt. Fitzroy, named for the captain of Darwin's *Beagle*. In Patagonia the peaks do not attain the heights of those farther north, but because they rise from a low plain of 2,000 feet, they are breathtaking. Fitzroy is a collection of bare granite needles climbing straight up, the tallest summit of which is eleven thousand feet above sea level. Glaciers curl about their roots. In order to climb the peaks, a number of years ago, an Italian mountaineering team spent days and nights dangling at the end of ropes like spiders.

Embarked on a much less ambitious adventure, we left our pleasant camp in a grove of trees encircled by a swift glacial river, finding our way by compass around Fitzroy's shoulder to the blue wall of its surrounding glacier. The bare gray needles reminded us of cathedral spires and at their feet we felt the same kind of awe as when one sees Chartres for the first time.

If the mountains of Patagonia are spectacular, so are its waterfalls. On one stretch of thirty miles we counted 125 cascades, some thundering torrents, other filmy bridal veils. Trees flourish in the damp climate of the mountains, and in remote areas we camped in primeval beech forests—nothofagus, or southern beech. To add to our delight, Patagonian parakeets flew by us in chattering flocks, while all around were masses of wildflowers, saxifrages, geraniums, and primulas. A condor, wondering what was going on in his isolated valley, flew low over us; with his eleven-foot wingspan he looked like an airliner.

One of the birds we particularly wanted to see was the Torrent Duck, a most rare creature, found only in this region. And sure enough, shooting down a series of rapids that would have daunted the toughest of kayakers, came a little duck which we swore had a smile on its beak.

One spot we'll never forget was a grassy terrace with a little nearby waterfall that Jane and I happened on. Off came our clothes, and under a fuchsia bush we lay in the cool water thinking that one could do worse than to come to Patagonia and take such a bath.   [16]

LIMITLESS ARGENTINE PAMPA

The plains of Nebraska, the Dakotas, Saskatchewan, seem to be without limits, to stretch on forever to the horizon. We believed nothing could exceed them—that is, until we ventured down to Patagonia and drove through the Argentine pampa. Such was the immensity of space and lack of landmarks that if the day was cloudy, we lost all sense of direction. It was like being at sea in a small boat. But featureless though the pampa was, there was always something to see; the clouds, the plants and the animals were all different to us from North America. Caracara eagles perched unconcernedly on fence posts as we drove slowly by on the abominable narrow gravel road leading south to Tierra del Fuego. Brilliant cerise flamingos waded in the infrequent swamps and guanacos galloped off to the horizon. Guanacos are the wild cousin of the tame llamas we had seen in Peru and are distantly related to the Near East's camel. Large, long-necked bird families scuttled through low bushes, mother and father leading a train of fourteen or so children trailing behind. They were rheas, this time a cousin of Africa's ostrich but having three toes instead of two.

Thousands of sheep wander slowly over the plain and in the lush bottomlands graze herds of fat cattle and beautiful horses, the Crillo horse, much prized as polo ponies but on the pampa used by Argentina's gauchos. We found that the gaucho is alive and well in Patagonia, seated on his sheep-skin-covered saddle, flatter than our western saddles. His lariat is attached to his right stirrup instead of to a saddle horn, and rather than wearing the cowboy's tight Levi's, his long, pleated dark baggy trousers are tucked into elegant low-heeled black boots. We never saw anyone use a bola—the three heavy balls attached to cords that, when thrown, entangle the legs of cattle and horses—but we saw them on sale in feed stores.

Because water is scarce on these high, dry plains, the amount of land needed to feed the animals is huge. We would drive all day and pass by only two estancias, as the ranches are called. Sometimes the signs at the entrance gates indicated that the main house might be twenty-five miles off the so-called main road.

Patagonia winds are ferocious, taking doors off cars, rolling our tents away across the pampa, windblasting the faces of the gauchos. Their bright red cheeks are the result of broken blood vessels; even the young children are thus afflicted. Small plants are deformed, and most bushes dare not stick their heads beyond three feet in height. Only the imported tall Lombardy poplars planted as windbreaks somehow manage to survive.

Flat the pampa may be, but to sleep out at night is an experience never to be forgotten, what with the bright stars of the Southern Cross so close you could almost reach up and touch its tail.   [17]

THE MYSTERIOUS MAYANS

Today the tangle of jungle creepers and thorns have long been cleared from the moldering lost cities of the Mayas. It is difficult to picture the massive monuments as they must have appeared to explorers John Lloyd Stephens and Frederick Catherwood in the 1840s. Now travelers at the Mayan sites see tidy lawns and restored buildings with the menacing jungle held at bay.

Both Stephens and Catherwood were well-qualified adventurers when they undertook in 1839 to look for mysterious cities in Central America. Catherwood, an artist, was particularly interested in Egyptian architecture, while Stephens, although trained as a lawyer, had written a successful book on his travels in the Near East. He was convinced after reading tales of early explorers that the cities really did exist in Central America although no one was sure who had built them. Antiquarians dismissed the Indians of the region as savages just emerging from barbarity and incapable of such work. The monuments must have come, they declared, from the hands of the Egyptians or the Phoenicians or the Lost Tribes of Israel.

The two men ventured into the interior of Honduras, enduring the dangers and hardships of the jungle, and found themselves after a number of hair-raising adventures amid the ruins of the huge city of Copan. They could scarcely believe what they saw. In the distance, almost covered with growth, was a vast stairway leading to a temple; about them stone jaguars reared on hind legs. Across an immense vine-choked plaza they saw a large, intricately carved stone figure in the jungle shadows. Stephens and Catherwood knew at once that although America was peopled by savages, savages had not built this city. But who were the people who had created this vast complex?

What race had reached such a height of sophistication as to be able to sculpt such figures? The two intrepid explorers, often wracked by malaria, pressed on, accompanied by a few machete-wielding Indians, to discover and clear all the major Mayan cities now accessible to modern travelers. They found that each site possessed artifacts, carvings, and architecture common to all and concluded that the ancestors of the Mayan Indians still living in the region had once created one of the greatest cultures of the New World.

When Stephens returned to New York to publish, in 1841, *Incidents of Travel in Yucatán*, illustrated by Catherwood's accurate and exquisite drawings, the world was astonished to realize that the ancient civilization of Central America was American and built by native Americans.

[18]

THE MAYANS PLAYED
A TOUGH GAME
OF BASKETBALL

Back when I was a teenager, I enjoyed my days on the basketball court. I still have a picture of the team and there I am in the center holding the ball, the captain. Little did I realize then what could have happened to a player in that position in Yucatán, Mexico.

Years later as I walked across the large ball field at the ruined Mayan city of Chichén Itzá in Yucatán I thought of my basketball-playing days. I don't think I realized back in my boyhood that a game very like the one developed in the early 1900s had been played by the Indians in Central America one thousand years ago.

The Mayan players certainly had a more impressive setting than any court I played on. Beautiful high stone walls rose on either side of an immense court that was 223 feet wide and 545 feet long! I thought of how exhausting it must have been to run around on it under Yucatán's fiery sun. At either end was a platform for the spectators. But what caught my eye, halfway along the length of the court, was a massive circular stone, about 20 feet from the ground, four feet in diameter, with a hole in its center. Across the court was another ring, exactly opposite to it.

In the game which they called pok-a-tok, the object was, of course, to get the six-inch-diameter hard rubber ball through the hole. But they made it hard for themselves; they could use only their fists, hips, or elbows; no dribbling, no throwing in Yucatán!

In the early days of the Mayan civilization the spectators kept an unusually keen eye on the game. If it looked as if the visitors were about to win, the onlookers ran away as quickly as they could, for the winning team had the right to strip the hometown rooters of all their finery.

As time went on and invaders from the north took over the Mayan cities, the rules got a lot tougher, in fact downright bloody. The captain of the losing team lost not only the game but his head. On one of the walls is a bas-relief of a kneeling captain, blood spurting from his neck, the winner holding a severed head. But did this action have religious significance? For the blood has taken the form of huge feathers. I wonder how I would have done as captain of a Mayan team. [19]

HUMAN NATURE

Nature is full of these wonders, dear cousin; we are admitted to the view of a very small section of it only; there is little hope then that we should be able to understand its relations fully, or to unravel all its mysteries.
— Jean Jacques Rousseau (1717–1779)

The search for the truth is in one way hard and in another easy—for it is evident that no one of us can ever master it fully, nor miss it wholly. Each one of us adds a little to our knowledge of nature and from all the facts assembled arises a certain grandeur.
— Aristotle (350 B.C.)

WHISKEY AND RUM

Spirits resulting from the distilling of fermented fruit or grain have been a staple part of the human diet at least since the earliest beginnings of cultivation some 8,000–10,000 years ago. Even in this new country, particularly during the boisterous century preceding the War for Independence and the more sobering period of early nationhood, spirits were as routine a part of daily life as that other staple and staff of life, bread. Even the Puritans, who railed against all fleshly pleasures, considered this daily nourishment indispensable and were in fact the first to distill rum on these shores.

Old-timers could not have survived without whiskey. Nothing equaled it as a disinfectant. All ages took it medicinally, from babies with colic and croup to the aged with rheumatism and tuberculosis. When no other remedies existed for the common cold or a chill, whiskey was their only other option, taken hot, sweetened sometimes with rock candy. Inflammations, including toothaches and infections, received liberal dousings, for many knew of little else to apply.

The ingenuity of pioneer life brought about some remarkable uses for spirits. The story is told in New Hampshire of a farmer who night after night lost his crops to bears. All his attempts to trap them failed. Finally he hit upon a somewhat extravagant, yet successful, foil. He and his son filled some small troughs with a mixture of rum and molasses, remembering bears' sweet tooth. These troughs they placed around the edges of the fields just at dusk. Early the next morning they found three staggering bears stumbling drunk through the fields and easily dispatched them.

In Vermont they tell the tale of the farmer who wanted to take his bull over to his neighbor's place so the bull could service his cows. Ornery as a mule, the bull would not budge. So the farmer emptied the better part of a bottle of whiskey into the bull's feed, and soon the farmer was able to lead the bull to his rendezvous.

For those who ran a homemade still, it was commonplace to keep pigs. Apparently the mash that remains after the distilling process can profitably be used to fatten pigs. Even one of the early farmers' almanacs in 1823 reported that feeding this mash to livestock fattens them better than whole grain does! [1]

CHEESE

Fuary, gary, nary,
Gary, nary, fuary,
Nary, fuary, gary!

If you are unlucky enough to be bitten by a dog you're afraid may be rabid, never fear: simply write this little rhyme on a piece of homemade cheese and feed it to the dog. Or so they used to do in Herefordshire, England. Whether the charmed cheese cured you as well as the dog remained to be seen.

Cheese has sustained human beings for many thousands of years. Goat- and sheep-milk cheese was the order of the day for early Greeks, but other folks regularly ate cheese made from yak, elk, buffalo, and reindeer milk. Pliny, the Roman naturalist who traveled the known world in the first century A.D., thought the sweetest cheese came from camel's milk. The Romans also had a hankering for smoked cheese.

Some of our early ancestors believed that the kind of milk given to young animals affected them significantly, so that once, when puppies and piglets were switched, the puppies brought up on sow's milk "oinked" like little pigs, while the piglets nurtured by the mother hound grew up to bark just like dogs.

Herb cheese, once very popular, was made with sage, tansy, pennyroyal, and balm into fat round green moons resembling the green cheese that makes the moon in the sky. The green in the cheese, however, usually came from spinach juice squeezed into the milk before it was curdled, so herb cheese commonly appeared in early summer, when the spinach was picked.

Thomas Jefferson received as an inauguration gift on New Year's Day 1802 one of the largest cheeses ever made, presented to him by Elder John Leland and the neighboring farmers of Cheshire in the Berkshires in Massachusetts. This colossal 1,450-pound cheese was made from milk given by their cows on a single day. It was brought in from all the farms and pressed in a cider press reinforced with hoops. Edwin Mitchel, in *It's an Old New England Custom* (1946), tells us they spent a month pressing the cheese before putting it to ripen in a cheese house, where it was carefully tended and delicately turned every day without cracking it. A pung—a horse-drawn, boxed-in sleigh—carried the mammoth cheese on the first leg of its journey from Connecticut to Hudson, New York, on the river, where a sailing ship took up the charge, transporting it to New York City. From there, six prancing horses decked with ribbons, drawing a colorful wagon, conveyed the huge cheese to Washington, where Thomas Jefferson cut the first slice with these words: "I will cause this auspicious event to be placed on the records of our nation, and it will ever shine amid its glorious archives."

[2]

SNEEZING

Hachoo! achoo! achoo!
Whatever will I do?
Sneeze before seven,
Company before eleven;
Sneeze before I eat,
Company before I sleep!

The nose is a door which lets the soul out and in—or the devil—whichever comes first, or so people all over the world have believed. In New Guinea, the Koita people are glad if someone sneezes while sleeping, for this is a sign that the person's soul is returning to the body. However, the ancient Persians thought sneezing meant an evil fiend was coming out of the body, and people around the sneezer uttered a prayer to keep the fiend from getting into them.

Saying "God bless you" harks back to days when the plague ravaged Europe. Sneezing apparently was a symptom of the plague, and people in Britain today still sometimes say, "God bless you, and may you not get the plague."

Do you like to make wishes? Whenever you sneeze once, this is the propitious time for wishing. Or did you know that if you sneeze twice, you will be kissed, perhaps by a fool? But sneezing three times means you will be disappointed. Four sneezes will deliver you a letter. However, five consecutive sneezes will bring you something new, while six sneezes indicates you will go on a journey. Yet among the Tonga, who live in the South Pacific, sneezing before a journey is a bad sign.

Do you sometimes feel someone is thinking about you? If you sneeze while having this thought, you will know it is true. Also, if you sneeze while trying to convince someone else of the truth of what you say, the sneeze proves you right.

Did you know the direction you sneeze is important? Sneezing to the right is a sign of prosperity, but sneezing to the left signifies you have cause to worry.

In Germany it is supposed to be bad luck to sneeze while

putting on your shoes. And in Estonia, we are told by Maria Leach in *The Dictionary of Folklore* (1950), "if two pregnant women sneeze together, they will have girls, but if two husbands sneeze, their children will be boys."

Now here's a rhyme people used to say about sneezing. Harry Hyatt collected this in Adams County, Illinois.

Sneeze on Monday for health.
Sneeze on Tuesday for wealth.
Sneeze on Wednesday for a letter.
Sneeze on Thursday for something better.
Sneeze on Friday for sorrow.
Sneeze on Saturday, see your sweetheart tomorrow.
Sneeze on Sunday, safety seek,
For the devil will be with you the rest of the week.   [3]

HICCUPS

Boo! We jump out at someone, hoping to scare them and drive away their hiccups. Or we blow up a paper bag and then clap it fiercely—bang! Hiccups go away. Surprise is the key element, catching the hiccupper off guard. Miraculously the hiccups are gone. When these noisy methods don't work, we instruct the sufferer to put her little fingers in her ears or to press the ends of her thumbs tight behind her ears and hold her breath or to stick her fingers in her ears while someone gives her a glass of water to drink. She may feel a little silly, but the result—no hiccups.

Water is reputed the essential element in many old-fashioned and new-fangled hiccup cures. Gulp down three or five or nine or twelve swallows of water, and the malady will be gone. Or bend over and drink water from the wrong side of the glass. Guaranteed to turn you into a contortionist, and some swear by it as an end-all for the hiccups.

Water is not the only potion recommended as a remedy

for this age-old, puzzling indisposition. Put a little sugar in a teaspoon of lemon juice, or soak a sugar lump in vinegar— surefire ways to pucker your mouth if not squelch your hiccups. Or try drinking a cup of water in which you have mixed some drops of peppermint. If these wet methods fail, old-timers turn to the certain dry method of putting your head in a paper bag and breathing your own exhaled air.

Perhaps everyone tries so readily to rid himself of hiccups because what the old folks believe really *is* true—that if you have hiccups, you have told a lie. Almost everywhere, hiccups are an embarrassment. But to some, if you get them, it may mean someone else is talking about you, or worse, that you are being attacked by the evil eye. In Illinois, if you have hiccups for a very long time, you are made to eat a potato three nights in succession instead of supper and then to take a spoonful of castor oil in hot coffee. Surely this ought to cure something. Should this not work, then the instruction is given to "lift up one end of a rock, but be careful to keep the other end on the ground, and then spit under the rock and restore the rock to its original position."

In Cornwall, England, out on the moors are three small, isolated standing stones, the center one round with a hole in the middle like a donut. A great variety of diseases are believed reversed if the afflicted person crawls three times sunwise through the hole.

Should you want to avoid having to go through any of these gyrations, including hopping on one foot and holding your breath or standing on your head, simply wear a nutmeg around your neck, and you will never get hiccups in the first place! [4]

STRESS MANAGEMENT

Hans Selye is known as the "father of stress." He didn't give birth to it or discover it. He is the person who did most of the early research on the subject. Selye defines stress as the wear and tear on our bodies, the aging process. A life without stress is impossible; neither is it desirable. Creative tension helps to bring out our best. It causes us to meet a challenge, grow, gain confidence, mature.

The stress that causes us problems is the stress that leads to *distress*, possibly disease, and eventually death. Stress can have lasting ill effects on our bodies. Selye calls causes of stress "stressors." He cites three kinds—physical, psychological, and social. Physical stressors include chemicals, pollutants, drugs, some foods, shock therapy. Psychological examples are intense emotions like hate, love, grief, self-pity, and especially frustration and anxiety. Social stressors are existence factors such as death, job changes, marriage, divorce, retirement—in other words, life events. Stressors, plus individual "make-up," determine the amount of stress, and this is modified (or increased) by the person's state at the time. "Make-up" is determined by one's heredity and environment.

While we cannot always control events or situations, we can control ourselves. We grow or decay, not by what happens to us, but by how we respond. We choose what we think, feel, say, and do. Managing stress means exactly that—instead of letting stress manage us. Coping is different things to different people, but there are guidelines to help us become a better-prepared person—relaxation, exercise, proper diet, short- and long-term goals, a good support network, an ability to play (have fun), and a value system. Self-awareness and deciding what kind of person you want to be help too. Are you a Type A or Type B? Which do you prefer for yourself?

A University of Kentucky bulletin has a good list of "ways to conquer stress":

- Concentrate on what you're doing
- Don't blame or complain
- Think of those you love
- Be "with" Mother Nature
- Run, sing, smile
- Give to others

- Exercise, meditate
- Learn to say no [5]

JOE RANGER

Joe Ranger used to stride down the road past our farm every two weeks or so, a jaunty little old gnome of a man. Over his shoulder he'd have a stick with a bandanna tied on the end. Joe was going for groceries down at the Four Corners. Not that he needed or used many store-bought goods; Joe was pretty self-sufficient. He lived up on Bunker Hill, named for the hill down in Massachusetts where the battle was fought. Vermont's Bunker Hill got its name because 216 years ago, folks living up there claimed they could hear the sound of cannons. Since that time the little settlement on top of the hill had disappeared, the road became a trail and only Joe Ranger lived there in a tumbledown house next to his beaver pond.

Sometimes in the summer we'd hike up to visit with Joe; he was a pleasant man and liked company in moderation. As I said, he was self-sufficient. He invited us inside his dwelling to show us his living arrangements. The only furniture was one chair in the middle of the room next to a wood stove. Everything was very efficient. He kept a pile of wood at hand, and around the edge of the room was his store of potatoes. When he wanted sweetening, he just reached in through a hole in the wall and fetched out a fistful of honey, courtesy of the bees who shared the house with him. All he needed to buy was some coffee and tinned meat. What kept him so spry, he claimed, was that he hadn't had to eat "no danged woman's cookin."

Sometimes hikers got lost and asked Joe where they were. He said that they would ask, "Where be we?" and he would reply, "Why you be right here."

His beavers gave him all the company he needed and

he liked to show how they responded to him. By the pond he would whistle and call, "Come, my dear little man, come here." From the other side a dark snout would push out of the water coming toward us, leaving a V in its wake. Then a small whiskery face would look up at Joe from the water's edge.

Joe didn't hold with doctoring. When he felt a little under the weather he soon set things to rights with a good dose of kerosene and porcupine blood. But the twentieth century finally caught up with Joe. Health officials worried about him and had him admitted to the hospital for a checkup. Joe died there. People in the neighborhood said that a bath probably killed him. He should have been left alone.

Joe's house has been bulldozed away, the beaver evicted, and rows of condominiums now surround a neat pond. The paved road is called Joe Ranger Drive.　　　[6]

RAINY-DAY HIKER

I have little use for people who shun the woods on a rainy day. They willfully deprive themselves of one of the better faces of nature. They leer while saying that a rainy-day hiker "don't know enough to come in out of the rain."

There is no time when the forest is more mysterious or more challenging—no time when a human being is more tempted to wonder where the animals and birds are keeping themselves. We can guess that deer are huddling under a grove of tall hemlocks and rabbits are snuggled into some leaf-covered brushpile, but we don't know for sure.

With a steady rain beating down through the trees onto half-frozen land, the only noise is heard in the rain itself. The dead, snow-packed leaves are soaked and quiet under the pressure of boots. A man can walk forever without disturbing the quiet.

In little depressions between the gently rising knolls of the forest, tiny rivulets will be running on their way to a newly formed stream, which in turn follows a natural course to the brook. Each stream is a novelty, because it will be gone at the end of the storm.

The tiny streams I see are in the watershed of Spoonerville Brook, and it takes hundreds of them to make the brook so full that it is, at the very moment that I watch on the hill a mile away, threatening to overrun its banks and flood the streamside dens of muskrat and mink.

A man walking the woods on a rainy day knows the makings of brooks, creeks, and rivers that get so rambunctious they break their ice cave and then threaten whole cities. Such currents are swollen from millions of trickles down the trunks of bare maples and birches.

He knows too there's nothing more peaceful than a day so rain-soaked that even the chickadees deem it wise to hunker down in an evergreen thicket and wait for the sun. Chickadees, it would appear, know when to come in out of the rain.   [7]

END OF THE TETHER

In 1845, the year he turned twenty-eight, things appeared to be slipping for Henry Thoreau. For a while after he graduated from Harvard in 1837 he taught school in Concord, but by mutual agreement he was dismissed after a mere two weeks because he could not cooperate with official school policies. Although the basic conflicts were deeper, Thoreau got himself into trouble with the local authorities because he refused to thrash his pupils.

Throughout the years Henry was writing. But apart from a few essays and articles in newspapers he had little published and had made little or no money. In the spring of

1843, at the urging of his friend and mentor Ralph Waldo Emerson, he went to Staten Island to try to establish himself in New York City. But that, too, failed and in the fall he was back home in Concord—never to leave again for any length of time.

That year Henry worked in the pencil factory; he also wrote in his journals, took his walks, and occasionally went on excursions with his friends. On one of them, a boating trip up the Sudbury River, he and a companion accidentally started a forest fire that burned over the precious woodlots of several local farmers. By this time Henry had a reputation in Concord—a Harvard man who had never yet held a regular job, a writer who did not earn money for his effort, a malingerer, eccentric, fire-starter, nature-lover. And then, as if to confirm his reputation, in the spring of 1845 Henry commenced the most rash act of his life.

For several years he had felt that what he needed to establish himself as a writer was time and space. He had thought about building a small cabin on the shores of Flint's Pond (now Sandy Pond) in Lincoln, but old man Flint, the so-called "owner" (as Henry would say) of the pond, had refused to allow him to live there. In hope of encouraging Henry in his writing career, Emerson offered to let him use some land he owned by the side of Walden Pond. In March Henry went to the shore, cut down a few pines, squared them off for timbers, raised the frame of a small cabin, sheathed it, roofed it, furnished it, and on the Fourth of July, 1845, moved in and subsequently entered what amounted to the most productive period of his life.

Two years and two books later Henry moved on, but he left behind a landmark. There is no body of water its size that is as well known as Walden Pond, neither in the Americas, nor in Europe, nor in the Far East. The pond, the man, the idea, are synonymous.   [8]

RECOLLECTIONS OF
A FARM WOMAN

She had her own hens. There was always bacon and eggs, fried eggs, fried potatoes, lots of hot oatmeal, and stacks of toast. Coffee. Sometimes she'd have pancakes and maple syrup. Sausage—her own sausage. Sometimes there'd be hash or beans. Doughnuts, and she served pie if you wanted it.

Farmers got up at four or five o'clock when they had to do all the work and the milking by hand. They'd go out and do two or three hours' work in the barn. By the time they came in, they were ready for a meal. So they ate breakfast then, about eight o'clock.

There was a big black cast-iron sink where the men washed up when they came in from the barn. You left your boots outside and ate in your stocking feet. There was a roller towel hanging beside the sink. There was a kind of carbolic soap for washing up—I think it was Lifebuoy. There was a cast-iron water pump on the end of the sink.

Yes, we burned candles when I was a kid. Used to make them ourselves. 'Course we had kerosene lamps too. We never had electricity. All our lights were tallow dips or kerosene lamps. You had to fill them up with kerosene and keep the wicks clipped. And wash the chimneys. There was one on each end of the dining room table when we were eating. And some of them hung up and had a reflector on the back. But we had candles.

Dances in a barn. Especially if someone had built a new barn, they always dedicated them with a barn dance. Square dances. There was a spring to the floor. And on those old square dances, you had to keep in time to the music and everybody else. If you didn't, you were going down when they were coming up—you met the floor. There was a dance hall in Ryegate Corners that was made that way—with a spring floor in it. I'll bet the floor would spring two inches.

You don't see too much of it nowadays, but there were

tramps around back then. I remember them going along the roads here in Hartland [Vermont]. Sometimes they'd stop and ask for something to eat—offer to cut wood or something to pay for it. Depression days and before that too. And there were the itinerants, the hoboes. Knew one named Tom Brown. He used to come to the farm and he'd work a while.

[9]

CLOG ALMANACS

Do any of our great-grandparents remember the clog almanacs that hung in the kitchen? Carved on the edges of a squared stick of hardwood, like a 2 × 2 about a foot long, the marks on the clog almanac represented not only the days of the week but all the special feast days, the cycle of the moon, and other important dates. Each of the four edges represented three months, or a quarter of the year. Thus the entire series of days constituting the year was notched along the angles of the squared wooden stick.

The first day of each month was shown by a longer stroke that turned up at one end. Each Sunday was distinguished by a broader stroke than the other days. A number of fascinating, pictograph-like signs became the tokens for special days—feast days, saints' days, and the like. St. Valentine's Day was marked by a lover's knot, the Welsh St. David's Day on March first by a harp. The hay harvest, which began traditionally on the eleventh of June, St. Barnaby's Day, was appropriately shown by a hay rake.

This custom of engraving a stick to make a calendar or almanac is very old. In fact, the people who lived in the Stone Age who painted the walls of their caves in Spain and France, the paleolithic hunters and gatherers of some 15,000–20,000 years ago, marked animal bones and deer antlers with long and short lines to keep track of the phases of the moon and thus to account for the passing of time.

These very early calendars suggest that our Stone Age ancestors held annual celebrations on fixed dates, which they kept track of by notching a tally on the bone, antler, or wood material convenient to them.

A clog almanac required no numbers or letters because all its marks are lines, dots, pictographs, or images. Such an almanac is really a perpetual calendar which shows the fixed holidays of the year.

In some places, particularly in Scotland, the special log burned at Christmas is still called the yule clog. We call wooden shoes clogs because we make them from blocks or clogs of wood.

Norwegians, Danes, and Saxons made special use of the clog almanac to engrave the periods of the full moon, new moon, quarter and half moons, regulating their festival days accordingly. However, knowing the phases of the moon is essential, as most farmers and gardeners know, if they are to plant and harvest at the most propitious times. Do they not remember, to ensure having sound cabbages, to set out the plants when the moon is one-third full? Or to plant trees when the moon is old so their roots will draw down into the earth?

For the activities that sustain life, such as those practiced by farmers, hunters, trappers, fishers, harvesters, and other people who live from the land and waters, knowing the phases of the moon is far more important than attaching a date or number to particular days. So long as the moon's cycle is clearly marked, tallied, accounted for, and followed, the pattern of life continues unbroken.   [10]

DWELLINGS

It is customary to think of the ancestral human dwelling as a cave—and half a million years ago some human beings did indeed live in caves—but in areas where caves didn't exist or in seasons and circumstances where caves were too confining or too stationary, primitive human beings also constructed windbreaks, lean-tos, and huts. Of these primitive structures, huts—being enclosed—are most like the houses we live in today.

The earliest huts were probably dome-shaped. Saplings or flexible branches were stuck into the ground in a circle and bent so that their tips could be tied together. This dome-shaped skeleton was then covered with grass, bark, leaves, or animal skins, depending on what was most available and most weatherproof.

Another type of primitive hut involved sinking both ends of flexible saplings or branches into the ground to create a series of parallel arches. These elongated, semi-cylindrical huts looked like miniature versions of the Quonset huts that the U.S. military introduced during World War II.

Interestingly enough, the European settlers who came to this continent during the 1600s encountered both of these primitive hut types in northeastern North America. The Algonquin Indians, who lived in what has become Massachusetts, built dome-shaped wigwams, and the Iroquois in what has become New York State built semi-cylindrical longhouses. But European architecture had evolved beyond such primitive shelters, and European settlers wanted semblances of their old homes in their new country.

The settlers came from many different architectural traditions, but timber and bricks had long since replaced the circles and arcs of primitive huts with the squares and rectangles of modern buildings. The preferred floor plan of colonial homes was rectangular because long side walls and short end walls provided the most efficient way to support a steeply pitched, weather-shedding roof.

In New England, timber-framed houses sided with clapboards and roofed with shingles predominated. Wood was the construction material of choice because there wasn't enough lime in the forested interior to make good mortar for laying bricks, and, furthermore, some of the early New Englanders believed bricks created an unhealthy environment to live in. Settlers farther south had no such reservations and, in Virginia, built themselves substantial brick houses to withstand the moister climate.

It took a group of Swedes who settled on Delaware Bay in 1638 to introduce the log cabin, which became the favored dwelling of the American pioneer, except in the prairies, where there were no trees. In the prairies, the sod house evolved, replacing the conical, wind-resistant, and easily transported teepees of the buffalo-following natives who preceded the European settlers.

In thinking about the many different forms that shelters have taken, I realized that the problem with modern houses and modern buildings in general is that they have evolved so far beyond merely sheltering us from undesirable weather, predators, and insects that they tempt us to stay indoors too much of the time. [11]

BLESS YOU, OLD BARN DOOR

There's an old barn door lingering in the distant past of many, keeping in the background of consciousness but never yielding its hold on visions of the opening and closing of a day's work. From a forgotten time, the rattle of its latch and the creak of its hinges call forth the best that is in us.

We think of it now and marvel at how much work its morning opening foretold and how much satisfaction its nighttime closing recalled. The cold complaint of its hinges at dawn would bring a low, expectant whinny from the horse stalls and the jingling of chains as cows rose to their feet, while its dropping latch at night locked in a hay-munching quiet.

Crude and simple, the door had to be tall enough for a horse and wide enough for a grass-laden Guernsey. It had to be strong enough to withstand the crush of two cows trying to enter at the same time and light enough so a pre-school boy could handle it without calling for help.

Barn doors have been made in as many designs as there are barns, with wood as the only common denominator and utility as the only common goal. We salute them, of whatever design, in their wheelbarrow-nicked ugliness and warmth.

The simplest barn door, built in a moneyless time, would have five eight-inch pine boards placed side-by-side with crossboards at top and bottom and a diagonal stiffener between. Unless the farmer was extravagant, the hinges and latch were removed from the battered and rotting door and installed on the new door. If the hinges were warped and past repair, old harness leather could be used as a replacement when there was no money for new steel.

Good times brought double-boarded doors with felt liners, put together with bolts instead of nails and bearing a coat of paint on the outside and whitewash on the inside. Better times brought rolling doors that jumped their tracks when slammed shut with too much boyish enthusiasm.

Whatever the design, opening the door in the morning signaled the start of a big day. It was opened seven days a week, year in and year out, with no sick leave, no holidays, no vacation, no money for extras like new hinges—nothing but the satisfaction of food and shelter for man and beast.

[12]

PLANETS & SPACE

KEEPING TABS
ON A WANDERING VENUS

The concept of morning and evening stars dates back to pre-telescopic days, when the ancients observed what looked like special stars that sometimes appeared above the eastern horizon just before the sun rose and sometimes appeared above the western horizon just after the sun set. These "stars" that seemed to have a special relationship with the sun were actually the planets Mercury and Venus at different phases of their revolutions around the sun, but the ancients perceived them merely as stellar wanderers and thought there were four of them.

Two of these wanderers were not very bright and never got very high above the horizon. These were Mercury, known in its morning appearances as Apollo and in its evening appearances as Mercury, which name it kept when it was recognized as a single planet. The other two were much brighter, sometimes achieving a brilliance fifteen times that of the brightest fixed star. These two were the planet Venus, known in its morning appearances as Phosphorus—torchbearer to Aurora—and in its evening appearances as Hesperus—leader of the nighttime stars.

Because Venus is much brighter than Mercury and climbs higher above the horizon, Venus is what most people see when they happen to notice a bright "star" appearing before the others in the evening sky. Likewise, a bright "star" that seems to shine on in the morning after the other stars have faded is most likely to be Venus at another stage of its long and leisurely cycle.

The reason Jupiter—and sometimes Mars and Saturn—are also referred to as morning or evening stars is that these other visible planets can sometimes shine brightly in the morning or evening sky. But they can also shine at other times of night, as opposed to Venus, which always precedes or follows the sun. Therefore Venus is the most characteristic morning or evening star and enjoys that designation in dictionaries and glossaries.

This business of morning and evening stars, which involves the wanderings of planets, is a bit more complicated than constellations, which for all intents and purposes stand still. Whereas the North Star is always in the same place—right above my barn—and the Big Dipper and Orion are always in the same locations with respect to the North Star, Venus sometimes appears on the eastern horizon in the morning, sometimes on the western horizon at night, and sometimes disappears altogether. [1]

CONTEMPLATING CONSTELLATIONS

Winter is stargazing season. The cold, clear nights provide the perfect conditions for viewing the stars and planets. If you can stand the cold for a few minutes, locate an open field and place yourself in the center of it. Look up and enter a spectacular otherworld.

There is good reason why winter is so good for watching the night sky. Think of our galaxy as a pancake. The earth is about three-quarters of the way out from the center of it. When we look at the Milky Way, we are actually looking sideways through the breadth of the pancake, rather than looking up and down through its depth, so there are many more stars to look through.

In winter, due to our revolution about the sun and our rotation on the earth's axis, night puts us, in the northern hemisphere, in the position of facing out toward the edge of the pancake. The stars we see are less dense because there are not as many in our line of vision as when we face inward toward the center. So what we do see is easily distinguished. The constellations are easier to determine because there is less of a clutter of stars from which to choose. Summer, on the other hand, has us facing inward toward the center of the galaxy, or pancake, so there are billions of bright and faint stars to confuse things. So if you want to begin learning about stars and the constellations they form, now is a good time.

To begin, there are five very good constellations to learn first. From these the rest of the sky will fall into place. These are the "circumpolar constellations," which are the ones that are visible all year long. They remain visible because they are located close to the North Star. The earth's axis is pointed almost directly at the North Star, so when the earth rotates on its axis once every twenty-four hours, the North Star stays in line. Anything just around the North Star will travel in a tight circle (so we see it), but things farther out from it will rise and set because the horizon gets in the way of our seeing the complete circle.

Think of it as if you were lying in a tent facing the roof. If the roof and walls were to spin around, you would be able to see the top center of it all the time and down the walls a little way. But, because you are facing in one direction, part of the sides will go behind you and then come back around from the other side. Circumpolar constellations are those that are painted at the top center of the sky roof. They spin around the very center point, the North Star, which never really moves; it just spins in place.

Many people are familiar with the circumpolar constellations without really knowing it. Because they are out all the time, they are the most commonly seen and easiest to learn. They are Ursa Major and Ursa Minor, the Big and Little Dippers, Cassiopeia (the queen), Cepheus (the king), and Draco (the dragon). These few constellations can be used as guides to lead you through the sky on a journey of wonderful discoveries. Science and myth come together in the stars like nowhere else. The best part is that this other-world is no farther away than our backyards.   [2]

METEORS

Comets have been held unlucky—at any rate, to kings and princes. But meteors—falling stars or shooting stars streaking across the sky and burning out in a very few seconds—are too transient to be threatening. However, in Ireland people used to believe that meteors were souls on their way from dead bodies to hell. Another more cheerful folk practice is to wish upon a star; but it must be done before the star burns out.

Meteors seem to be the result of the dispersal of the heads of comets formed into streams, and different meteor streams have their orbit around the earth. These orbits may be intersected by the earth's orbit, and, when our gravitational pull captures some of the meteors, the captives write their lines of light across our atmosphere, tiny scraps of ore burning out.

Meteor showers occur regularly. Two of the times of year when our orbit intersects with meteor streams are summer and winter. The winter show is called the Geminids, from a part of the sky with the constellation Gemini in the background. This is around December 9 to 14. Another shower—but fainter—is the Perseids, from the constellation Perseus. This shower occurs between the first and twentieth of August. Halfway through this period, the careful watcher may see as many as fifty or sixty in an hour.

Meteors burn out; but meteorites make it through our atmosphere to land on the earth—luckily not too often. The meteor crater near Winslow, Arizona, is almost a mile in diameter, but since it is probably ten thousand years old its damage to human populations was slight. Meteorites have been known for ages, but their origin in the heavens wasn't known until the early 1800s. When a fifty-four-pound meteor fell in Yorkshire, England, in 1795, people thought it had been hurled through the air from the volcano Mt. Hecla in Iceland, then in eruption.

There seems to be a great deal to occupy our minds these days, but fortunately death by a falling meteorite isn't one of them.   [3]

UNDERSTANDING
THE AURORA

The aurora, also known as the northern lights, is light coming from high above the earth that can be seen during the night in the polar regions. At times, the aurora appears to be only a few feet above our heads, but actually it is never less than fifty miles (eighty kilometers) above the earth. People used to believe that the aurora was the sun's light reflected from ice on the Arctic Ocean or from ice crystals in the atmosphere. Today we know that the aurora is actually light created where we see it in the sky. A typical aurora may be sixty or seventy miles above the earth and several hundred miles long.

The sun is a ball of very hot gases. These gases are so hot that some of the atoms blow away as part of what we call the "solar wind." Electrons stripped from the outer shells of these atoms also make up much of the solar wind. It takes the solar wind about three days to travel the ninety-three million miles between the sun and the earth. When the electrons reach the earth, they are trapped by the earth's magnetic field. They travel along magnetic field lines in a corkscrew fashion into the earth's atmosphere, where they emit light, creating the aurora.

The earth has a magnetic field that is shaped as if a large bar magnet ran through the center of the earth. The magnetic field controls the location of the aurora by guiding electrons into the earth's atmosphere near the geomagnetic poles.

These tiny particles can bounce back and forth along the magnetic field lines from the north to the south in a few seconds, creating similar and simultaneous auroras in the north and south polar areas.

Yes, there is an aurora in both polar areas. Most of the time the aurora in the south and the aurora in the north are exactly alike and move around in the same way at the same time. The northern aurora is "aurora borealis," meaning dawn of the north. The southern aurora is "aurora australis," meaning dawn of the south. [4]

CLOUDS

Look up in the sky. Can you tell from the clouds what the weather will be tomorrow or the next day? Most of us know when it's going to rain, but only just before it starts to pour. People haven't always been so out of touch with the sky. Long before there were radio or TV weather reports, or even daily newspapers, people who worked outdoors knew how to read changing cloud patterns. It's something anyone can do. All you need to learn is how to recognize the most common cloud shapes and what they mean.

The fun begins when you start predicting the weather.

Clouds are formed by water vapor that rises up into the atmosphere. The water vapor comes from Earth's oceans and lakes, but also from the plants and animals that inhabit the land. Water vapor is invisible. You can't see it until it cools enough to return to its liquid form. The water vapor is then said to condense, meaning it forms droplets of water. These droplets cling to tiny pieces of dust, smoke, or minerals that are floating in the air. A cloud forms when large numbers of these water droplets (or in some cases ice crystals) come together.

Although the same types of clouds occur the world over,

not all clouds are alike. Imagine how boring it would be if they were! Luckily, clouds come in a number of different shapes, sizes, and colors, depending on how high up and how thick they are.

The height of the clouds you see is one signal of approaching weather. The other is the direction in which the wind is blowing. While not always accurate, the old rhyme

> *Red sky at morning, sailors take warning,*
> *Red sky at night, sailors' delight.*

describes the fact that storms typically blow in from the west. But not all storms come from that direction. Here are some general guidelines to help you recognize changing weather: You can expect a storm if high, scattered clouds get thicker, increase in number, and get lower. If, however, low, dense clouds rise higher and decrease in number, or if a dense layer of clouds wrinkles up, thins out, and lets patches of sky show through, fair weather is on the way.   [5]

EVER WONDER WHY THERE'S WIND?

We have wind because the earth has an atmosphere full of air that is acted upon by the sun. When the sun heats the earth, the air above that portion of the earth is heated by radiation. The warm air rises and starts drifting toward a cooler place. Meanwhile, cooler air descends and races along the earth's surface to replace the warm air that has risen—creating wind. Because the amount of air in the atmosphere is constant and temperatures are different at different points on the earth's surface, air is always in motion, warm and cool trying to achieve an equilibrium they will never find.

One basic pattern of air flow is from the hot equator toward the cold poles and back again, but not all the hot air moving high through the Northern Hemisphere makes it straight to the North Pole. At about 30° latitude, some of the hot air has cooled enough to descend to the earth. A portion of this air finds itself invited back to the equator to get heated again, while another portion continues northward, traveling along the earth's surface.

This northerly traveling air is deflected by the earth's rotation and blows across areas between 30° and 60° latitude—which includes most of the United States—as what are called the prevailing westerlies. These prevailing westerlies bring us much of our local weather, blowing it from Chicago to New England in about a day's time. But there are other influences that keep wind and weather from ever being that simple.

For instance, some of the traveling air develops into self-contained systems that are held together by a swirling motion and are blown along by the prevailing winds. Some of these systems are full of cool air sinking heavily toward the earth, where they create areas of high pressure. Other systems are full of warm air rising from the earth to create areas of low pressure. These swirling highs and lows cause different winds locally, the highs causing light winds and the lows causing strong winds.

In addition to these high- and low-pressure systems, huge air masses form in the heat of the tropics or in the cold at the poles. These tropical and polar air masses float around in the atmosphere like invisible flying saucers, bumping into each other and creating what meteorologists call fronts. These collisions cause turbulent winds wherever they occur.

The final set of influences that shape the exact winds in any area are bodies of water, mountains, valleys, buildings, trees, nights, days, and seasons.   [6]

EARTHQUAKES—
ADJUSTING TO STRESSES

Earthquakes are not simple, formulaic, easily predictable phenomena. Nor are they chance, random, totally unpredictable accidents. They exist in that frustrating middle area where scientists can quantify what has happened in the past but make only educated guesses at what might happen in the future.

Earthquakes in the western part of the United States are somewhat easier to comprehend than earthquakes in the East. In the West, two continental plates—the North American and the Pacific—are in direct contact, and when points of contact along this highly stressed boundary slip, California has an earthquake. The eastern part of the United States is right in the middle of the North American plate, however, so the earthquake that shook us on October 7, 1983—a 5.2 on the Richter scale, which ranges from 0 to 8.9, increasing tenfold as it moves from number to number—cannot be explained by shifting plates.

The best scientists have been able to do with eastern earthquakes is to plot all the earthquakes that have occurred and theorize about the geological conditions that caused them. One theory holds that partially molten material beneath the earth's crust produces waves, which, when they hit a fracture in the rocks of the crust, cause an earthquake. Another theory holds that eastern earthquakes are most likely to occur near plutons—intrusions of younger rock into older rock. As the younger rock cools, it contracts and eventually wrenches itself free from the older rock, causing an earthquake. A third theory is that the Northeast is still rebounding after the last glacier, and underground readjustments cause our earthquakes.

Right now earthquake research is deeply involved in recording, measuring, and analyzing the earthquakes that occur and more cautiously involved in predicting the earthquakes that might occur in the future. Seismographic stations all over the world record quivers, tremors, major quakes, and disasters, and by comparing data establish the focus, epicenter, and magnitude of each particular disturbance. The focus is the underground point at which the earthquake originates, while the epicenter is the surface point directly above the focus. The magnitude is the amount of energy the earthquake releases.

From the focus, various waves, which we perceive merely as shakings and quakings but which seismologists can distinguish by their timing and behavior, radiate at measurable rates. Three sets of waves—primary, secondary, and surface—travel at different speeds and respond to different kinds of rocks in different ways. Seismologists have learned to interpret all the information that registers itself on their very sensitive instruments in ways that allow them not only to pinpoint the focus, epicenter, and magnitude of the earthquake but also to analyze some of the underground structures that lie between the focus and the seismographic station. [7]

SUMMER CLOUDS

As long as the clouds are so prevalent in this summer's weather, it's a good time to review the differences among them. Typical on a sunny summer day are the puffy white clouds which accumulate in late morning and early afternoon, causing trouble only for frustrated sunbathers. To understand how and why cumulus (from the Latin word meaning heap) clouds form helps explain how all clouds form.

All air contains a certain amount of moisture—whether or not you notice it depends on two factors: the amount of moisture and the coolness of the air temperature. The more moisture in the air, the less cooling it takes to produce a cloud. Conversely, the drier the air, the colder it has to become before clouds are visible. When invisible, the moisture is actually in a gaseous state, with the droplets too tiny to reflect light and therefore too tiny to see. "A cloud," according to Eric Sloane's *Weather Book*, "is the air's moisture gone from invisible gas to visible water droplets."

On a warm summer day, air, warmed by the earth, rises skyward and takes moisture with it. As more and more moisture rises, the air becomes unable to absorb all the vapor, which then condenses into tiny droplets of water, forming a cloud. Also, as air distances itself from the warm earth, it cools, lessening its ability to contain moisture vapor. On any given day, the clouds will form at the elevation where the temperature is cool enough to cause the invisible water vapor to condense into visible water droplets. The temperature is referred to as the dewpoint. Similarly, a glass of ice water chills the air surrounding it to the point where moisture in the air condenses on the outer surface of the glass.

Whereas cumulus clouds are puffy and heap up like cotton candy, stratus clouds are horizontally flattened and layered. When high in the sky, they may hint at an approaching storm; when low and dark, they are the clouds of extended rain. The highest clouds (around 31,000 feet) are cirrus clouds, composed not of water droplets but, in those frigid temperatures, of ice particles. A sky webbed with high, icy wisps of clouds may be the first sign that warmer, wetter weather is approaching. Often the sun or moon shining through sunshine particles will cause a rainbow halo around itself.

The wonderful variety of clouds in the summer reflects the great contrasts in temperature between earth and sky and the availability of water which lies frozen through the winter months. Although we're all ready to see a few cloudless days, it can be fascinating to watch the changing shapes and colors of summer clouds. [8]

WEATHER LORE

The north wind doth blow,
And we shall have snow.

Mares' tails and mackerel scales
Make lofty ships furl their sails.

If snow begins at mid of day
Expect a foot of it to lay.
When the snow falls dry,
It means to lie,
But the flakes light and soft
Bring rain oft.

When the wind is in the east
'Tis neither good for man nor beast.

When there is a circle around the moon, count the stars within the circle. They will tell you how many days will pass before the storm comes.

Two full moons in a calendar month (the second is called a "blue moon") will bring a large storm or a flood.

If the moon is old, the snow will likely last.

The white frosts and next a storm.

When snow comes with the new moon, it will melt quickly.

When the moon is low in the south during February, it means thirty days of good weather.

Warm February, bad hay crop.
Cold February, good hay crop.

A summerish January, a winterish spring.

If there's ice in November that will bear a duck,
There'll be nothing after but sludge and muck.

If October brings heavy frost and winds, then will January
and February be mild.

Dry August and warm
Doth harvest no harm.

Dog days bright and clear
Indicate a good year,
But when accompanied by rain,
We hope for better days in vain.

Calm weather in June
Sets corn in tune.   [9]

THE ENVIRONMENT

& CONSERVATION

GAIA: THE EARTH

We call our planet the Earth. We call the ground under our feet earth. By this simple act of language we create a confusion which we have only begun to unravel since J. E. Lovelock's book *Gaia: A New Look at Life on Earth* (Oxford, 1982) gave us a new name for our planet, Gaia. The primordial earth goddess for the Greeks had this name Gaia. By giving this name to our planet, we do more than merely distinguish the whole biosphere from its soil. By calling our planet Gaia, the name of an ancient goddess, we recognize that our planet is an ancient living system, is sacred, and is our mother.

Scientists are slowly beginning to recognize that this world of our planet Gaia is "a sensory, perceptual, psychological power," a living biosphere. We no longer can think of ourselves as living "on" the planet, but rather as living "in" or "within" it. Even when we sit on the earth in the summer shade of a maple tree in full leaf, we sit "in" the air we breathe. This air, this atmosphere, is the real skin of our planet. Yet because the breathed air is invisible, we often disregard its importance in our lives, in all life. Air is the medium we live in as fish live in water. We live "within" the outer layer of the living biosphere of our planet, breathing it into our lungs, breathing its oxygen into our blood, and thus breathing ourselves alive.

With every breath we affect this atmosphere that surrounds and supports us and that sustains and maintains our life. As every schoolchild knows, with our every breath we draw oxygen out of the air, depleting its supply, and replace it with carbon dioxide. When we sit under the summer maple tree, we trade chemical substances with the tree, giving the tree our unwanted carbon dioxide, which the tree breathes, and in turn receiving from the tree its discarded oxygen. This exchange, even expressed in these simple terms, is a metaphor for the deep communication, the communion between all the animal, vegetable, and mineral beings that live "in" the planet, in Gaia.

We live in the womb of the Great Primordial Earth Mother Gaia as completely as a baby lives in its human mother's womb, utterly dependent for all its needs on the health and wholeness of its mother. We need Gaia to be unpolluted because her air is our breath and our blood; her water is the living fluid in our bodies, which are 97 percent water; her soil is the ground of our being. Gaia is the only world we know, the only world we can know. Gaia is the world that encompasses us, that we are immediately "of" and "in." [1]

THE MANAGER
OF THE BIOSPHERE

Every human society creates and celebrates mythic heroes. They serve, in part, to illuminate proper behavior and reinforce the moral purpose of the group. They help to educate each new generation. We have traditionally honored the cowboy, the self-made man, and many other variations on the theme of rugged individualism. Such archetypes influence virtually all dimensions of a society's public and private life, including the basic relationships with the rest of nature.

Our tiny planet Earth deserves a new mythic hero—one who reflects our modern sense of the close interdependence of all living things, who willingly assumes the high calling of stewardship, who struggles to maintain a global vision in the face of parochial, short-sighted, and excessive demands on nature, and who sees our entire living space as a domain for enlightened management. We ourselves—the flawed, contentious, ever-expanding members of the human family—collectively constitute the only qualified candidacy for this inspiring role.

We have already caused continental ice caps to recede, oceans to rise, deserts to bloom, forests to vanish, and species to disappear. We have intruded directly into genetic processes and changed the course of evolution. We have eradicated smallpox and given birth to new cardiovascular afflictions. We have reversed the flows of some rivers, obstructed the flows of many others, and caused a few to ignite with industrial residues. We threaten biology itself with our weapons of war, and we deploy our technology to sustain lives that would otherwise have expired. We use seeds as political weapons and diminish wild areas that generate variety, resilience, and renewal. We are fruitful, and we multiply—often at a terrible cost to the rest of nature.

In other words, we are already the managers of the biosphere—however inept, partial, and destructive our management. We are effectively in charge of our own life-support system. But now we urgently need a new vision of ourselves that matches our new responsibilities, a new awareness of the human being as an inspiring mythic figure who affects the universe through responsible decision and positive action. Such self-awareness may lead, first, to better management of the biosphere and, later, to an appreciation of the human enterprise as a more dignified endeavor than we had previously imagined. The intelligent future of life on this planet is too vital to do otherwise.   [2]

THE END OF LIVING AND THE BEGINNING OF SURVIVAL

The following is an excerpt from a speech delivered in 1854 by Chief Seattle of the Duwamish tribe to the newly arrived Commissioner of

Indian Affairs for the Washington Territory. Portions of this text
have been rewritten by professor Ted Perry of Middlebury College.
The speech itself—eloquent, emotional, and painfully insightful—
is, unfortunately, as strong an indictment today as it was in 1854.
It is a remarkably prophetic statement of current environmental
ethics, presaging even John Muir by nearly half a century.

The Great Chief in Washington sends words that he
wishes to buy our land. The Great Chief also sends us words
of friendship and goodwill. This is kind of him, since we
know he has little need of our friendship in return. But we
will consider your offer. For we know that if we do not sell,
the white man may come with guns and take our land.

How can you buy or sell the sky, the warmth of the
land? The idea is strange to us. If we do not own the freshness
of the air and the sparkle of the water, how can you buy
them?

Every part of this earth is sacred to my people. Every
shining pine needle, every sandy shore, every mist in the
dark woods, every clearing and humming insect is holy in
the memory and experience of my people. The sap which
courses through the trees carries the memories of the red
man.

The white man's dead forget the country of their birth
when they go to walk among the stars. Our dead never forget
this beautiful earth, for it is the mother of the red man. We
are part of the earth, and it is part of us. The perfumed
flowers are our sisters; the deer, the horse, the great eagle—
these are our brothers. The rocky crests, the juices of the
meadows, the body heat of the pony, and man—all belong
to the same family. So when the Great Chief in Washington
sends word that he wishes to buy our land, he asks much
of us.

The Great Chief sends word he will reserve us a place
so that we can live comfortably to ourselves. He will be our
father, and we will be his children. So we will consider your
offer to buy our land. But it will not be easy. For this land is
sacred to us. This shining water that moves in the streams
and rivers is not just water but the blood of our ancestors. If

we sell you land, you must remember that it is sacred and that each ghostly reflection in the clear water of the lakes tells of events and memories in the life of the people. The water's murmur is the voice of my father's father.

The rivers are our brothers; they quench our thirst. The rivers carry our canoes and feed our children. If we sell you our land, you must remember, and teach your children, that the rivers are our brothers and yours, and you must henceforth give the rivers the kindness you would give any brother.

The red man has always retreated before the advancing white man, as the mist of the mountains runs before the morning sun. But the ashes of our fathers are sacred. Their graves are holy ground, and so these hills, these trees, this portion of the earth is consecrated to us. We know that the white man does not understand our ways. One portion of land is the same to him as the next, for he is a stranger who comes in the night and takes from the land whatever he needs. The earth is not his brother, but his enemy, and when he has conquered it, he moves on. He leaves his fathers' graves behind, and he does not care. He kidnaps the earth from his children. He does not care. His fathers' graves and his children's birthright are forgotten. He treats his mother the earth and his brother the sky as things to be bought and plundered, sold like sheep or bright beads. His appetite will devour the earth and leave behind only a desert.   [3]

THE TROPICAL EQUATION

We have all seen or heard versions of the statistics—that 4 square miles of tropical rain forest might contain as many as 1,500 species of flowering plants, 750 species of trees, 125 species of mammals, 400 species of birds, 100 species of reptiles, 60 species of amphibians, and 150 species of butterflies. No one even knows how many species of other insects there might be, although a report from the National Academy of Sciences has estimated that as many as 42,000 insect species may reside in just 2.5 acres of typical rain forest. No other ecosystem on earth contains greater diversity.

We have seen and heard other kinds of statistics. The United States' National Academy of Sciences puts the total figure of felled primary tropical forest as 124,000 square miles each year. At current rates of deforestation, nearly all tropical forests will have been eradicated from the planet within fifty years.

Nor is it just the rain forests that are disappearing from the tropical equation—so are wetlands and shorelands and coral reefs. And with them, the glorious mix of life-forms they support.

We worry a lot about such figures here in the United States. Organizations are formed to protest the destruction of tropical habitat and wildlife in places like Brazil, Malaysia, and Africa, where agriculture, logging, cattle-raising, and industrial development combine in a continuing assault on the natural world. Books and articles and television documentaries expose such destruction to the harsh and honest glare of journalism, and editorials chastise the peoples and governments of countries presumably less enlightened than those of the industrial nations. And we should worry. The world should worry. The loss of so much life, so much of the richness and health and beauty that once blessed this planet above all others, is a tragedy whose final implications cannot even be guessed at with any authority.

But while we worry, however legitimately, however necessarily, we in the United States would do well to take a long hard look at the tropical inheritance of life over which *this* nation and its people have assumed stewardship. There is more of that special web of life in our hands than we normally realize—we have our own rain forests, our own fragile island ecosystems, our own coral reefs. And the careless energies we have loosed on such places as the Florida Keys, the Caribbean islands, the southern Rio Grande Valley, and the Hawaiian Islands do not seem to provide this nation with much moral ground on which to stand when it pre-

sumes to tell others what to do. In this case, as in so many
others, extinction begins at home—and it should end here,
too.   [4]

THE UNPAYABLE COSTS
OF GLOBAL WARMING

Forward-thinking people are preparing for the greenhouse
effect. Barge companies on the Mississippi are acquiring rail-
roads in case the river becomes permanently unnavigable.
Planners of Boston's sewage-treatment system are taking
into account a sea-level rise from global warming. The Wey-
erhaeuser Company is planting drought-resistant trees. The
Dutch are raising their dikes. Either the behavior is certifiably
crazy or it's an apt assessment of the craziness of the human
race as a whole—a reasoned bet that we will be stupid
enough to let the greenhouse effect happen instead of stopping it.

   I'm not that much of a pessimist. I think we're rational
animals, at least when it comes to economics. The economics
of the greenhouse effect are quite clear. We can't afford it.

   The Environmental Protection Agency (EPA) has com-
missioned studies of the effects of a projected global climate
change on the United States. In unemotional bureaucratic
language, the report spells out unmitigated disaster. Wheat
and corn production may shift away from the Great Plains.
The agricultural economy may no longer be able to sustain
the rural population. Under the driest scenario, projections
for the Great Lakes region and New England are that species
like eastern hemlock and sugar maple could disappear. Mature natural forests in the region could be reduced from one-
quarter to one-half their present acreage, with many poor
sites giving way to grassland or scrub conditions. There will
probably be disruptions and/or reductions in the availability

of major forest resources—wood, water, wildlife, recreation opportunities.

The United States could lose up to 70 percent of its coastal wetland with a one-meter rise in sea level. A one-meter rise would inundate an area the size of Massachusetts. Most of these losses would be concentrated in the Southeast, particularly Louisiana and Florida. The EPA says that coastal areas now in hundred-year floodplains would in fact be subject to storm surges an average of every fifteen years. Hurricanes would form more often and be stronger.

Salt water would invade groundwater aquifers, and the salt-fresh interface would move higher up river mouths—endangering water supplies from Cape Cod to New York, Miami, and California's Central Valley. Protecting coastal cities with bulkheads, levees, and pumping systems would cost $30–$100 billion. Raising barrier islands by pumping sand onto them would cost $50–$100 billion. (That would double property taxes for those islands' residents, says the EPA, but that's better than losing their property altogether.) We will probably have to gradually remove structures from much of our coastal lowlands, says the agency.   [5]

VERMONT FORESTS WEATHER THE ACID TEST

An explosion of dead and dying trees threatens to tarnish Vermont's reputation as the Green Mountain State. Many believe that acid rain and other air pollutants are the culprits behind the devastation of some of the northeastern forest ecosystems. "The damage at higher elevations is absolutely staggering," says botany professor Dr. Hubert Vogelmann, a researcher and leading authority on acid rain. "It can't yet be scientifically proven that acid rain is the primary cause of forest decline, but the mounting evidence links air pollution directly to it."

Vermont has suffered drastic changes over the last thirty years. Some tree species are losing their vigor and dying, and 80 percent of Vermont's red spruces already have disappeared from the mountain forests. The amount of standing wood at higher elevations is down 44 percent since 1965.

The Experimental Station's acid rain lab is breaking new ground by focusing its research on forest decline. "Mountain forests are in a more advanced stage of damage than are those in lower elevations because clouds and fog bathe the mountains in high doses of acid and other pollutants," Vogelmann says. "We've studied Camel's Hump Mountain, trying to form a blueprint of the damage that could possibly spread elsewhere."

The acid rain lab is co-directed by Vogelmann and Dr. Richard Klein. Klein, who is a plant physiologist, conducts laboratory experiments, while Vogelmann is more involved with field studies. The acid rain lab has gained worldwide attention by showing a strong link between air pollution and forest decline. Klein's precise experiments, in particular, have demonstrated how acid rain and harmful pollutants in the environment can weaken trees.

Camel's Hump and Hub Vogelmann have become associated with acid rain, both having been mentioned in major newspapers around the world. Scientists, politicians, and the media come from all over to see the mountain and observe the researchers' operation. The Experimental Station's success also has spurred countless other studies and millions of dollars of funding for further research.

Vogelmann gives lectures to organizations and universities about once a month, hoping his findings will awaken people to what acid rain is doing to their natural resources. "Scientists have a social responsibility to clearly explain to the public what's happening and what can be done about acid rain," Vogelmann says. "After all, it's the public that is ultimately making all of the decisions, so it has to know what's going on."

Acid rain research has come a long way, but there still are no concrete answers or solutions. Vogelmann says, "We're looking towards the future, studying what's happening to our forests and what they will look like in the years ahead. Vermont has done the pioneering studies, though, to help solve the problem of acid rain." [6]

GEORGE PERKINS MARSH

Over a hundred years ago we were warned of our wasteful ways by a Vermont lawyer, George Perkins Marsh. In a remarkable book called *Man and Nature: Physical Geography as Modified by Human Action*, published in 1864, Marsh foretold our problems and developed a scientific study to deal with them—ecology, the concept of the interrelationship between organisms and the environment.

When the first Europeans came to North America, the continent was rich in natural resources, rich beyond the wildest dreams. Clear fish-filled streams, primeval forests, grass as far as the eye could see, an earth filled with minerals—all was here for the taking. The frontiersman, the farmer, the business entrepreneur never doubted that such bounties would last forever.

In *Man and Nature* Marsh anticipated the crisis in pollution, overpopulation, and vanishing resources and stated that man was no asset to the earth. "Even now," said Marsh, "we are breaking up the floor and the wainscotting and doors and window frames of our dwelling to warm our bodies and seethe our pottage." These prophetic words were written in 1864. Today we know that we have recklessly let acres of fertile crop and range land the size of France wash or blow away or have allowed it to disappear under suburban sprawl. The cost of cleaning up our polluted environment is enormous, but most frightening to our twentieth-century civilization is the realization that the sources of inexpensive energy, on which that civilization is based, are drying up at an alarming rate. We have indeed used our capital of natural resources in a most careless and thoughtless manner.

George Perkins Marsh was without doubt one of the

most extraordinary of beings, a man of boundless enthusiasms and massive intelligence. In his long life he managed to write a definitive book on the origin of the English language, a Scandinavian grammar, and to form one of the first collections of art in America. He was also a bankrupt businessman who at the age of fifty-five regarded himself as a failure. He headed the commission that designed the present Vermont statehouse in Montpelier when the old one burned down in 1857. He helped to found the Smithsonian Institution. The final design of the Washington Monument was his. He wrote a book urging the use of camels in the American desert. He was an expert instrument maker, a lawyer, a politician, a master of twenty languages, a minister to Turkey, and finally, the first American minister to the newly formed Kingdom of Italy, a post he held for twenty-one years, a record unsurpassed by any other American diplomat. But of far greater importance, he was this country's first environmentalist.   [7]

WE NEED NOTHING
SHORT OF A REVOLUTION

Few of us picking up trash and carrying Save the Earth banners are thinking of a thoroughgoing revision of human culture. But down deep most of us know that's the direction our concern will lead. We know it is impossible to go on finding and wasting oil, leveling forests, paving land, dumping poisons, and multiplying our numbers. A new way of life, a new set of thoughts *must* be found.

A new thought is already here, in fact, in the idea of sustainability, a word that has just surfaced in the discourse of the industrial world. It means, very simply, using the planet's resources in a way that doesn't diminish the resources of future generations.

It's as impossible for us to describe a sustainable world as it would have been for the farmers of 6,000 B.C. to foresee present-day Iowa or the English coal miners of 750 to imagine a Toyota assembly line. We know only that such a world would use energy with painstaking efficiency. It would respect, reuse, and recycle materials. It would have a stable population. The thoughts in people's heads would be about harmony with nature rather than conquest.

Like the other great revolutions, an environmental revolution will require sacrifices and lead to enormous gains. It too will change the face of the land and human institutions, hierarchies, self-definitions, and cultures. It will take centuries.

That is, if it happens. There is no guarantee. The alternative is the ecological and economic impoverishment of a culture that cannot adapt to its environmental necessities.

William D. Ruckelshaus, former Environmental Protection Agency administrator, expressed in a recent *Scientific American* the size of the challenge: "Can we move nations and people in the direction of sustainability? Such a move would be a modification of society comparable in scale to only two other changes: the agricultural revolution of the late Neolithic and the industrial revolution of the past two centuries. Those revolutions were gradual, spontaneous, and largely unconscious. This one will have to be a fully conscious operation, guided by the best foresight that science can provide. . . . If we actually do it, the undertaking will be absolutely unique in humanity's stay on the Earth." [8]

IT WILL TAKE TIME
TO PERFECT RECYCLING

Contrary to conventional wisdom, Americans are proving not only able but willing to separate their garbage. As the market for used newsprint crashes, however, some recyclers are getting discouraged, and some purveyors of conventional wisdom are saying, "See there? Recycling just doesn't work. There's no market for it." It would be more accurate to say we don't know if recycling works; we haven't yet tried it. When we do, gluts and scarcities will be signals not that there's no market, but that the market is working the way it always works—in fits and starts.

What we are doing so far is separating, not recycling. We're beginning to reclaim materials before they get to the dump. We have barely begun to close the loop—to re-use major materials in the same products: newspapers back to newspapers, plastic soda bottles back to soda bottles. Product-to-same-product recycling is the only kind that can work in the long run. Turning newspapers into cattle bedding will be helpful for a while, but eventually it will clog, either because we use newspapers faster than cattle bedding or vice versa. Similarly, the plastics industry is congratulating itself too soon for turning soda bottles into plastic flowerpots. Given the nation's consumption rate of soda versus flowerpots, one can easily predict a market collapse due to flowerpot glut. What works is illustrated by the nation's one smooth-running and economical recycling system—aluminum cans back into aluminum cans.

Even when newspapers are printed on recycled newsprint and the plastics industry makes new bottles out of old, there will be glitches, scarcities, and overflows. These are inevitable in the evolution of any production system, especially one that is guided by the market. The only way the market can sense a large potential supply of something new is to let that something accumulate somewhere. The only way the market can stimulate a demand is to bring the price down low enough and be sufficiently assuring about future supply to stimulate new users. In short, don't let a temporary newspaper glut discourage you. We're just at the beginning of a major industrial transformation. We're working out a material-supply system consistent with a finite planet. It will be totally different from the wasteful, polluting system we have now—and it will take a while to get it right.

It's worth keeping part of our attention cast ahead of the

immediate economics to the place where we're ultimately headed. A sustainable, economic, ecologically supportable materials system will re-use everything it can. It will add virgin materials only as necessary to sustain product quality. It won't waste materials on unproductive purposes such as packaging—it will use uniform and minimal packaging, standard bottles or boxes of standard sizes, interchangeable among products, for easy re-use. Only the label will distinguish the product. Marketers will have to attract consumers with a reason to buy their product. That's more important than glitzy packaging. [9]

DON'T TAKE IT ANYMORE

Environmentalists are often criticized for focusing on the negative, telling people what they shouldn't do without offering alternatives. Solid waste is one area in which we all contribute to the problem but can also become part of the solution. Massachusetts generates approximately 6.1 million tons of solid waste per year—that's about one ton per person, or seven pounds of garbage per person per day. Even more startling, these figures continue to increase every year, as the percentage of packaging, products, and other garbage that is nonrecyclable continues to increase. The main reason we are finding more and more plastics, polystyrene, and other nonbiodegradable products and packaging is that they are cheaper at the checkout counter. If the price you paid in the store reflected the ultimate cost of disposing of nonrecyclable materials, then throwaway products would certainly be more expensive than they are now. Until market prices reflect this hidden cost of disposal, we will continue to see a decrease in environmentally sound packaging. The following lists a few ways to reduce contributions to the waste stream.

• Choose paper cups, plates, or other disposable ta-

blewear over plastic or Styrofoam. Better yet, avoid
disposable products whenever possible.
- In the checkout line, ask for paper bags instead of
quietly accepting plastic ones.
- If you have the option of choosing between two brands
of a product, buy the one with less packaging; reward
manufacturers that are environmentally aware.
- Buy rechargeable batteries, reusable razors, returnable
milk bottles—all of these products conserve resources
and energy.

Consumers must demand alternatives. The cola companies didn't invent caffeine-free cola because they were worried about customers with the jitters; cola-free was created
because there was a consumer demand. The same principle
works for products and packaging; as soon as industry recog-
nizes that environmentally sound products can be marketed
as such, you will find them at the corner market. If your
grocery store still provides paper shopping bags, let them
know you appreciate the service. How much more would
paper cups cost the coffee club in your office? Do you return
the bottles and cans that have bottle deposits? Let retailers,
corporations, and manufacturers know why you prefer certain products over others.   [10]

HOUSEHOLD WASTE DISPOSAL

While you are shopping consider the products you are buying; a product is not safe simply because it can be bought in
a store. Many products in your home may be hazardous—
these include home or garden products, paints, batteries,
and other common items that contain dangerous chemicals.
Many communities are now holding "household hazardous
waste collection days" so that unused products can be collected and disposed of safely. Old paint, pesticides, and clean-

ing solutions often end up in landfills, where they have the potential to contaminate groundwater. Whenever possible, substitute nontoxic alternatives to the "traditional" drain cleaners, pesticides, or insecticides found in the store.

Automotive products can be especially dangerous: one gallon of gasoline can contaminate 750,000 gallons of drinking water. Used automotive oil contains heavy metals, chlorinated solvents, and organic compounds. It is estimated that 8 million gallons of used oil are generated each year by passenger vehicles in Massachusetts. State law requires that all stores that sell oil also accept used oil for recycling, but participation by "do-it-yourselfers" is very limited. If you change your own oil, be sure to recycle it. Oil that is poured on the ground or down the drain ends up in our lakes, streams, groundwater, and harbors.

Disposal costs of at least $65 per ton to landfill solid waste are common. In the autumn up to 40 percent of the waste stream is leaf and yard waste. Many towns now offer municipal composting operations and actually derive income from selling compost as mulch. Even if a town doesn't make money selling compost, taxpayers save money for every ton of organic material that is kept out of landfills. If you live in the suburbs or country, consider starting your own compost pile. Most basic gardening books explain the steps necessary to maintain a safe, odor-free composting bin in the backyard.

[11]

INDOOR AIR QUALITY

Researchers have known for decades that indoor air can be polluted in hidden ways. But only in the past few years have builders, homeowners, and health officials become concerned with the problem.

Combustion, or burning, is one of the most obvious sources of indoor air pollution. This can mean anything from a poorly installed woodstove to a steak sizzling on the kitchen stove. In most cases, though, we can reduce our exposure to contaminants by controlling or adjusting their sources.

Unvented kerosene and gas space heaters are particularly bad sources of harmful combustion gases. Because these heaters release gases directly into the indoor air, they should be used only as emergency heat sources. The best way to deal with air-quality problems from unvented space heaters is to simply not use them.

Gas stoves and ovens are not as hazardous as unvented kerosene heaters, but they too can release undesirable quantities of nitrogen dioxide, carbon dioxide, and carbon monoxide into your home. Using an exhaust fan or opening a window near the stove when you cook will usually vent these gases.

Smoking is a well-known source of several hazardous air pollutants, the most severe being respirable suspended particulates (RSPs). RSPs affect not only smokers, but others near them. This is a difficult pollution source to deal with because it involves the rights of consenting and nonconsenting participants. Whether the smokers are strangers or loved ones, removing the source of smoke is seldom a simple matter.

Vented heating devices such as furnaces and boilers usually don't contribute to indoor air pollution, but serious problems can result if a flue is blocked or broken. Proper insulation and an annual heating system inspection, cleaning, and tune-up can ensure against most problems. Wood- and coal-burning stoves release combustion by-products when they're loaded or when wind conditions cause a downdraft. This problem is accentuated in tightly sealed buildings because air infiltration for combustion is restricted. Opening vents to supply outside air to combustion sources can alleviate this problem.   [12]

SICK BUILDING SYNDROME

The typical American lives day in and day out with hundreds, perhaps thousands, of toxic chemicals. A recent study focused on the levels of just twenty or so volatile organic chemicals in indoor and outdoor air and in people's breath. The investigators found homes two to five times more polluted than outdoors, and in isolated cases, indoor air was as much as a thousand times more polluted. This held true even in New Jersey, the chemical state, where the outdoor air is heavily burdened. The investigators found that indoor air was sufficiently polluted to raise the risk of acute systemic reactions (such as nausea, headaches, fatigue, and confusion), cancer, and birth defects.

In modern office buildings, low levels of chemical irritants combine with inadequate fresh-air intake and low humidity to produce "sick (tight) building syndrome." Formerly attributed to mass hysteria, its symptoms include sleepiness, nausea, headaches, and eye-nose-throat-sinus irritations. Almost everyone in these buildings is affected to some extent. Perhaps 30 to 40 percent are distinctly uncomfortable; a few become intensely ill and possibly hypersensitive.

In many workplaces, especially in new buildings, exposures to dangerous chemicals are far greater than in the usual office. In one, monitored by Environmental Protection Agency (EPA) investigators over a three-month period, scientists identified several hundred volatile organics and nearly two dozen known mutagens and carcinogens. The high level of toxicity was attributed to paint, sheetrock, wallpaper, carpeting, glue, cleaners, and insecticides.

Certain workplaces are at least as hazardous as new buildings, and some are orders of magnitude more so. Among workers singled out by EPA researchers as having especially high levels of toxics in breath samples were those working in dry-cleaning establishments, chemical industries, plastics, textiles, painting, metalwork, wood processing, gas stations, laboratories, printing, dye plants, and hospitals.

Tens of millions of American workers (how many precisely is not known) are exposed to toxics in doses high enough to compromise their health. Lance Wallace, principal author of the TEAM study, fears that the incidence of chemical sensitivity may be rising sharply. "I get a lot of calls from people who believe they're sensitive," he says. "It certainly seems to be a problem."   [13]

DRY HOMES

An often-heard complaint in the middle of the winter concerns the dryness of the air indoors. Symptoms of a dry house range from a dried-out nose and mouth when you wake up in the morning to the legs falling off chairs. People with wood heat seem to have more problems than those with other types of heating systems. All of this is a result of conditions that exist during cold weather.

Cold air can hold less moisture than warmer air. On a cold winter morning the air temperature may be −10°. The relative humidity might be 65 percent. While the relative humidity sounds high, in reality very little moisture is in the air because it is so cold. If this air is brought into the house and warmed up, the amount of moisture in the air remains the same, but the relative humidity goes down, possibly as low as 5 or 10 percent.

The air, brought in from outside, is desert dry. The dry air now absorbs moisture from any place it can around the house—your skin, the furniture, the dog's dish, or any other source. If the house is very leaky, this warmed air that has collected moisture from around the house leaks outside and is replaced by more dry outside air. The process goes on and on, and the house gets drier and drier.

If you heat with wood or coal, outside air constantly flows into the house for combustion. It is this continual flow of air that dries everything out. If you have an oil- or gas-fired heating unit, outside air still flows into the house for combustion, but it is not continuous. Thus the moisture loss is less. So wood heat is not really drier than other kinds of heat. It's the air flow that's different.

An activity that goes on in many homes at this time of year involves adding moisture to the air. This is done by placing pans of water on the stove or registers, venting the clothes dryer inside, drying clothes inside on lines, or even running humidifiers. There are as many ways of adding moisture to the home as there are people to think them up. In all cases, some trade-offs have to be considered.

As moisture is added to the air, the relative humidity goes up. As this moist air comes in contact with cold surfaces, the moisture condenses. This condensation first appears on windows. It may occur on walls or ceilings, especially in corners or behind furniture. The condensation may be in the form of water, or in very cold weather, it will appear as frost.

The trick is to add enough moisture to the air to achieve comfort, but not so much as to cause condensation problems. With wide fluctuation in the outside temperature, this balance becomes a real challenge. If the moisture levels are brought up during the day, condensation problems may occur at night, when both the outside and inside temperatures go down.

No magic solution exists for the problem of moisture and cold weather. Adding moisture to the air to increase comfort has to be balanced against condensation problems. Each house will be different as to this balance point. As the outside temperature changes, so will the balance point. The only sure thing is that the problems will go away when spring gets here.   [14]

WOOD SMOKE

Wood smoke contains numerous toxic and irritating agents that can be deposited in the deepest regions of the lungs, where they damage the tissue. Although there has been limited research about how wood smoke affects public health, studies have shown that wood smoke contains seventeen of the Environmental Protection Agency's (EPA) priority pollutants, fourteen carcinogenic compounds, and toxic gases such as carbon monoxide.

Wood emits 17 times the particulate matter, 550 times the carbon monoxide, and 150 times the benzo-a-pyrene (a carcinogen) emitted by oil for a given amount of heat. A smoldering fire, a full stove, and wet wood—conditions that promote incomplete combustion—increase pollution. The practice of loading up the stove and closing the damper to get slow, even heat thus increases the production of harmful gases and pollutants.

Recognizing that wood smoke pollution is a national problem, the EPA established emission standards for wood stoves sold after July 1988. Many of these models have catalytic converters or are specially designed to reduce pollution.

If you do not have a new stove, there are a number of commonsense steps you can take to get the most heat and the least amount of airborne pollutants from your wood stove.

- Since green wood produces fires with lower temperatures and higher amounts of pollutants, burn wood which has been seasoned for six to eight months.
- Make a fire and let it burn briskly for half an hour. Most air pollution occurs within the first thirty minutes, before the stove has heated up.
- Load the stove more often with smaller amounts of wood. Large fuel loads decrease the circulation of air as well as the firebox temperature. Open the draft for 10–30 minutes whenever you add a load of wood.
- Use the size of the fire rather than the damper to regulate heat.
- Don't burn trash, treated wood, or plastics, all of which may produce harmful emissions.
- Use a stack thermometer. Your stove is burning efficiently and producing the least amount of pollutants when the flue gases are 300–400°F. You can also

monitor your chimney. An air-starved fire will produce smoke, while a hot, clean-burning fire will not.

[15]

WHERE THERE'S FIRE, THERE'S SMOKE

For me it has been most satisfying over the years to step out
to the woodpile and survey the orderly stacks, savoring the
satisfaction of having plenty of seasoned wood to last until
spring. These are not mere logs, but stacked memories of
crisp fall days in the woods, of tangy smells of oil and freshly
cut maple, of the hearty pleasures of honest, sweaty work.

Now, I understand if you are a resident of Missoula,
Montana, and a light mounted on top of the city's water
tower is flashing to signify an air-pollution alert, you could
be cited for lighting or continuing to feed any wood fire.
Missoula may have gone a bit farther than other municipali-
ties, but it is not alone in taking an increasingly jaundiced
view of a pastime that is still regarded as an American birthright—like whittling, hunting, or driving enormous cars.

This year the government imposed stiff new restrictions
on manufacturers of wood stoves. And more than one city
has considered ordering its residents to seal their fireplaces.
Just a decade and a half ago, when the first Arab oil embargo
tripled petroleum prices, created all-day gas station lines and
produced electric bills the size of mortgage payments, the
wood fire took on the status of organic gardening and newspaper recycling as a thinking person's way to save both
money and the environment.

American industry, delighted to be able to offer something for sale that did not trigger protest marches on Washington, responded with the airtight stove, which was soon
selling at the rate of 2 million a year. When oil prices dropped

again, the frenetic pace slowed to about 500,000 per year, but by the mid-1980s there were an estimated 12 million wood stoves in use throughout the United States. At least another 12 million American households had working fireplaces.

The smoke from a few widely scattered rural wood stoves or the fireplaces of a Norman Rockwell village adds a spicy tang to dead winter air. But crowd hundreds of thousands of fires into a densely populated city—especially one surrounded by mountains and shrouded by an atmospheric inversion—and the pleasant flavoring, concentrated, becomes highly toxic. This all-American product produces carbon monoxide gas and—especially from airtight stoves throttled down to burn slowly—microscopic particles of incompletely burned wood known as polycyclic organic matter, many of which are classified as carcinogenic by the Environmental Protection Agency. [16]

OUR FATE DEPENDS . . .

A thing we are just learning is that both the genetic library and the ecosystem's services depend on the integrity of the entire biological world. All species fit together in an intricate, interdependent, self-sustaining whole. Rips in the biological fabric tend to run. Gaps cause things to fall apart in unexpected ways.

For example, songbirds that eat summer insects in North America are declining because of deforestation in their Central American wintering grounds. European forests are more vulnerable to acid rain than American forests because they are human-managed single-species plantations rather than natural mixtures of many species forming an interknit, resilient system.

Biodiversity cannot be maintained by protecting a few favorite species in a zoo. Nor by preserving a few greenbelts or even large national parks. Biodiversity can maintain itself, however, without human attention or expense, without zoo-keepers, park rangers, foresters, or refrigerated gene banks. All it needs is to be left alone.

It is not being left alone, of course, which is why biological impoverishment has become a problem of global dimensions. There is hardly a place left on earth where people do not log, pave, spray, drain, flood, graze, fish, plow, burn, drill, spill, or dump.

Biologist Paul Ehrlich estimates that human beings usurp, directly or indirectly, about 40 percent of each year's total biological production (and our population is on its way to another doubling in forty years). There is no biome, with the possible exception of the deep ocean, that we are not degrading. In poor countries biodiversity is being nickeled and dimed to death; in rich countries it is being billion-dollared to death.

To provide their priceless service to us the honeybees ask only that we stop saturating the landscape with poisons, stop paving the meadows and verges where bee food grows, and leave them enough honey to get through the winter.

To maintain our planet and our lives, the other species have similar requests, all of which add up to: Control your-selves. Control your numbers. Control your greed. See your-selves as what you are, part of an interdependent biological community, the most intelligent part, though you don't often act that way. *Act that way*. Do so either out of a moral respect for something wonderful that you did not create and do not understand, or out of a practical interest in your own survival. [17]

IN THE END

In your Bible, you might be familiar with the account, at the opening of the Book of Genesis, describing how the world began. Here is a different version:

In the end, there was Earth, and it was with form and beauty. And man dwelt upon the lands of the earth, the meadows and trees, and he said, "Let us build our dwellings in this place of beauty." And he built cities and covered the Earth with concrete and steel. And the meadows were gone.

And man said, "It is good."

On the second day, man looked upon the waters of the Earth. And man said, "Let us put our wastes in the water that the dirt will be washed away," and man did. And the waters became polluted and foul in their smell.

And man said, "It is good."

On the third day, man looked upon the forests of the Earth and saw they were beautiful. And man said, "Let us cut the timber for our homes and grind the wood for our use." And man did. And the land became barren and the trees were gone.

And man said, "It is good."

On the fourth day, man saw that animals were in abundance and ran in the fields and played in the sun. And man said, "Let us cage these animals for our amusement." And there were no more animals on the face of the Earth.

And man said, "It is good."

On the fifth day man breathed the air of the Earth. And man said, "Let us dispose of our wastes into the air for the winds shall blow them away." And man did. And the air became filled with the smoke and the fumes could not be blown away. And the air became heavy with dust and choked and burned.

And man said, "It is good."

On the sixth day man saw himself; and seeing the many languages and tongues, he feared and hated. And man said, "Let us build great machines and destroy them lest they destroy us." And man built great machines and the Earth was fired with the rage of great wars.

And man said, "It is good."

On the seventh day man rested from his labors and the Earth was still, for man no longer dwelt upon the face.

And it was good. [18]

Contributors

THE SEASONS

1. Spring Walk. Jane Curtis.
2. The Mad March Hare. Eleanor Ott, Vermont Folklife Center and the Center for Research on Vermont, January 1985.
3. Calvin Coolidge's Glorious Fourth. Jane and Will Curtis and Frank Lieberman, *Return to These Hills: The Vermont Years of Calvin Coolidge*, Curtis-Lieberman Books. Reprinted by permission of the Calvin Coolidge Memorial Foundation, Inc.
4. The Icebox. Will Curtis, December 3, 1985.
5. Haying. W.H. Bunting, *Sanctuary*, May/June 1984, Massachusetts Audubon Society.
6. Fall. Wayne Hanley, *Nature's Ways*, No. 760, October 1978, Massachusetts Audubon Society.
7. Fall Color—No Mystery at All. Linda Garrett, Vermont Institute of Natural Science Nature Column, October 2, 1989.
8. Nuts! *The Drummer*, December 1988, The Ruffed Grouse Society.
9. Christmas Eve. Eleanor Ott, Vermont Folklife Center and Center for Research on Vermont, December 1984.
10. The Christmas Tree. Gale Lawrence, *Rutland Herald/Times Argus*, November 1982.
11. Wreaths. Information provided by Vermont Folklife Center, University of Vermont Center for Research on Vermont, November 1984.
12. St. Stephen's Day. Information provided by Vermont Folklife Center, University of Vermont Center for Research on Vermont, November 1984.
13. The Mistletoe Tradition. Gale Lawrence, *Rutland Herald/Times Argus*, November 1979.
14. Snow-Season Survivors. Kim Fadiman, *Sierra* magazine, February 1984.

PLANTS, GARDENS, & TREES

1. Wildflowers: Miracles of Spring. Lee Goodman, *Biologue*, Spring 1989, Teton Science School.

2. Red Osier Brings Color to March. Gale Lawrence, *Rutland Herald/ Times Argus*, March 1980.
3. Spring Flowers. Bonnie Ross, Vermont Institute of Natural Science Nature Column, April 28, 1986.
4. Spring in the Garden. Jane Curtis.
5. Seed Dispersal: Ingenious Ways to Get Away. Jenepher Lingelbach, editor, *Hands-On Nature,* Vermont Institute of Natural Science, 1986.
6. The Growth of a Seed. Bonnie Ross, Vermont Institute of Natural Science, May 10, 1985.
7. Soil. Dolores Savignano, *Sanctuary*, April 1983, Massachusetts Audubon Society.
8. Planting Trees. Gary Moll, *American Forests*, March/April 1990.
9. Leafy Wonders. Maywin Thoreson, *South Dakota Conservation Digest*, Vol. 56, No. 1, 1989.
10. Papyrus. Naphali Lewis. Reprinted with permission of *Archaeology* magazine, Vol. 436, No. 4. Copyright © 1983 by The Archaeological Institute of America.
11. Grasses: Slender Stalks with Seeds that Nourish the World. Jenepher Lingelbach, editor, *Hands-On Nature*, Vermont Institute of Natural Science, 1986.
12. Dandelions: Survivors in a Challenging World. Jenepher Lingelbach, editor, *Hands-On Nature*. Vermont Institute of Natural Science, 1986.
13. Gotcha! Jim Gordon, *Sierra* magazine, October 1986.
14. Silage. *New England Farm Bulletin*, May 8–21, 1981.
15. Fascinating Facts of Fungi. Rick Youst, *Biologue*, Fall 1986, Teton Science School.
16. The Morel Mushroom. Dave Ode, *South Dakota Conservation Digest*, Vol. 57, No. 2, 1990.
17. Lichens. Vermont Institute of Natural Science.
18. In Defense of Deadwood, Decadence, and Disorder. Ted Williams, *Sanctuary*, January 1981, Massachusetts Audubon Society.
19. Galls: Small Homes for Tiny Creatures. Jenepher Lingelbach, editor, *Hands-On Nature*, Vermont Institute of Natural Science, 1986.
20. Mangroves. Carrol B. Fleming, *Américas*, March/April 1983.
21. Balm of Gilead. Eleanor Ott, Vermont Folklife Center and Center for Research on Vermont, December 1986.
22. For What Ails You. Megan Parker, Vermont Institute of Natural Science, March 13, 1989.
23. The Tree of Life. John H. Mitchell, *Sanctuary*, No. 3, September 1989, Massachusetts Audubon Society.
24. Talking Trees. Eleanor Ott, Vermont Folklife Center and Center for Research on Vermont, December 1986.
25. Catnip. Eleanor Ott, Vermont Folklife Center and Center for Research on Vermont, January 1987.
26. The Cattail. Dave Ode, *South Dakota Conservation Digest*, Vol. 55, No. 5, 1988.

27. Herb Ideas. Leonard Perry, University of Vermont Extension Service, May 20, 1988.
28. Preparing for a Soil Test. Lee Reich. Reprinted courtesy of *Horticulture, The Magazine of American Gardening*, 20 Park Plaza, Suite 1220, Boston, Massachusetts 02116. Copyright © 1990, Horticulture Partners.
29. White Birches: Two Varieties. Gale Lawrence, *Rutland Herald/Times Argus*, September 21, 1986.
30. Autumn's Hidden Harvest. Robert H. Boyle. Copyright © 1987 by the National Wildlife Federation. Reprinted from *National Wildlife* magazine, October/November 1987.
31. Sumac. Richard Headstrom, Vermont Institute of Natural Science Newsletter, October 1981.
32. All About Pumpkins. Monica Porter, University of Vermont Extension Service, October 26, 1981.
33. When Plants Mark Time. Ned Smith. Copyright © 1971 by the National Wildlife Federation. Reprinted from *National Wildlife* magazine, February/March 1971.
34. Watering Houseplants. Leonard Perry, *Green Mountain Gardener*, University of Vermont Extension Service.
35. Cacti. Leonard Perry, *Green Mountain Gardener*, University of Vermont Extension Service, December 25, 1987.
36. Specialist Offers Tips on Buying Fuelwood. Lisa Halvorsen, University of Vermont Extension Service, November 28, 1986.
37. Genetic Improvement of Sugar Maple. Howard Kriebel, Ohio Agricultural Research and Development Center, The Ohio State University.
38. Why Does the Sap Run? Gale Lawrence, *Rutland Herald/Times Argus*, March 4, 1986.

ANIMALS

1. Keeping Warm: Five-Dog Night. John H. Mitchell, *Sanctuary*, December 1983, Massachusetts Audubon Society.
2. Messages in the Snow. George H. Harrison. Copyright © 1989 by the National Wildlife Federation. Reprinted from *National Wildlife* magazine, February/March 1989.
3. Communication. Jenepher Lingelbach, Vermont Institute of Natural Science Newsletter, January/February 1984.
4. Camouflage. *The Drummer*, February 1989, The Ruffed Grouse Society.
5. Color It Survival. George H. Harrison. Copyright © 1984 by the National Wildlife Federation. Reprinted from *International Wildlife* magazine, July/August 1984.
6. A Look at Animal Vision. Melissa Abramovitz, *Sierra* magazine, March/April 1989.
7. The Brumation of Ectotherms. Tom Tyning, *Sanctuary*, October 1989, Massachusetts Audubon Society.
8. Snakes Maximize Success with Minimal Equipment. Adrian Forsyth, *Smithsonian*.

9. Night Flyers. Thom Engel, *The Conservationist*, May/June 1989, magazine of New York State's Department of Environmental Conservation.
10. Earthworm. Richard Headstrom, Vermont Institute of Natural Science Newsletter, August 1978.
11. Mink. Wayne Hanley, April 1979, Massachusetts Audubon Society.
12. The Gray Squirrel. Bill Vogt. Copyright © 1988 by the National Wildlife Federation. Reprinted from *National Wildlife* magazine, August/September 1988.
13. A Look at House Mice. Gale Lawrence, *Rutland Herald/Times Argus*, July 21, 1985.
14. The Porcupine Proves Its Point. Gary Turbak. Copyright © 1986 by the National Wildlife Federation. Reprinted from *National Wildlife* magazine, October/November 1986.
15. Where Have All the Woodchucks Gone? Bonnie Ross, Vermont Institute of Natural Science, October 17, 1986.
16. Skunks. Jenepher Brettel, Vermont Institute of Natural Science, February 26, 1981.
17. Red and Gray Foxes. Linda M. Garrett, Vermont Institute of Natural Science Newsletter, Winter 1988–89.
18. White-Tailed Deer: Adapting to All Seasons. Jenepher Lingelbach, editor, *Hands-On Nature*. Vermont Institute of Natural Science, 1986.
19. Lyme Disease Affects Both Humans and Animals. Lisa Halvorsen, University of Vermont Extension Service, April 28, 1989.
20. Limits to Growth. Elaine Tietjen. *Habitat: Journal of the Maine Audubon Society*.
21. Coyote. Thomas Conuel, *Sanctuary*, December 1988, Massachusetts Audubon Society.
22. A Wolf Somewhere Sings. Bruce Thompson, "Looking at the Wolf," Teton Science School.
23. Black Bears. Nancy Bell, *Vermont Environmental Report*, Spring 1990, published by the Vermont Natural Resources Council, 9 Bailey Avenue, Montpelier, Vermont 05602.
24. Polar Bears. William J. Kelly, editor, *Seaword*, Vol. 17, No. 4 (Fourth Quarter 1989). Mystic Marinelife Aquarium, Sea Research Foundation, Inc., Mystic, Connecticut.
25. Cattle Country. William H. Bunting, *Sanctuary*, No. 5, May/June 1989, Massachusetts Audubon Society.
26. Sugarelle. John H. Mitchell, *Sanctuary*, May/June 1989, Massachusetts Audubon Society.
27. Miracle at Midnight. Barney Crosier, *Rutland Herald/Times Argus*, June 1980.
28. Training Oxen and Steers. Reprinted from *Yankee Drover: Being the Unpretending Life of Asa Sheldon, Farmer, Trader, and Working Man, 1788–1870*, foreword by John Seelye, by permission of University Press of New England.
29. Camels. James C. Simmon, *Audubon*, January 1991.

BIRDS

1. Bird Legends and Lore. Donna J. P. Crossman, Vermont Institute of Natural Science Newsletter, Fall 1984.
2. But for the Grace of—Feathers? Susan Diamond, Vermont Institute of Natural Science Nature Column, November 1989.
3. Birds and their Reflections. Wayne Hanley, May 14, 1980, Massachusetts Audubon Society.
4. Bird Nests. Jane and Will Curtis, *Backyard Bird Habitat*, published in 1988 by Countryman Press, Woodstock, Vermont.
5. Nests, Nestlings, and Chicks. Katy Duffy, *Biologue*, Spring 1989, Teton Science School.
6. The Molt. Nancy L. Martin, Vermont Institute of Natural Science Nature Column, August 1985.
7. The Smartest Family of Birds? Leah Barash. Copyright © 1989 by the National Wildlife Federation. Reprinted from *National Wildlife* magazine, October/November 1989.
8. Nasal Notes and Headfirst Hops. Jenepher R. Lingelbach, Vermont Institute of Natural Science Nature Column, March 3, 1986.
9. The Right Tools for the Job. Claire Miller. Copyright © 1989 by the National Wildlife Federation. Reprinted from *Ranger Rick*, August 1989.
10. The Vanishing Barn Owls. Rick Mooney. Copyright © 1988 by the National Wildlife Federation. Reprinted from *National Wildlife* magazine, June/July 1988.
11. Raptors: Another Cog in Nature's Wheel. Eileen Dowd, *South Dakota Conservation Digest*, Vol. 55, No. 5, 1988.
12. Terns. Massachusetts Audubon Society, 1988.
13. Outrageous Fortune. Dr. Danny Ingold and Mr. Daniel Otis. Reprinted from *The Living Bird Quarterly*. Summer 1989, a publication of the Cornell Laboratory of Ornithology.
14. Bald Eagles. Pat Cole, The Yellowstone Association Newsletter, Summer/Fall 1989, Vol. 5, No. 3.
15. Peace Symbol. Donna Johnson. Copyright © 1991 by the National Wildlife Federation. Reprinted from *National Wildlife* magazine, December/January 1991.
16. Goldfinches. Wayne Hanley, August 1980, Massachusetts Audubon Society.
17. Hawks. Wayne Hanley, Massachusetts Audubon Society.
18. The Black-Capped Chickadee. Richard Headstrom, Vermont Institute of Natural Science Newsletter, January 1978.

INSECTS

1. Secret Life of a Forest. George H. Harrison. Copyright © 1987 by the National Wildlife Federation. Reprinted from *National Wildlife* magazine, June/July 1987.

2. Insects: Amazingly Adapted and Adaptable. Jenepher Lingelbach, editor, *Hands-On Nature*. Vermont Institute of Natural Science, 1986.
3. Katydids. James L. Castner. Copyright © 1989 by the National Wildlife Federation. Reprinted from *International Wildlife* magazine, March/April 1989.
4. Crickets. Mary Holland Richards, Vermont Institute of Natural Science.
5. Mosquitoes. Lisa Halvorsen, *Creature Feature*, July 14, 1989, University of Vermont Extension Service.
6. Houseflies. *The Indoor Naturalist*, copyright © 1986 by Gale Lawrence. Published by Prentice Hall Press, a division of Simon & Schuster, New York, New York.
7. Fireflies. Wayne Hanley, August 1979, Massachusetts Audubon Society.
8. Honey. Dadant and Sons, Inc., Hamilton, Illinois, October 29, 1990.
9. Femme Fatale. Charles Roberts. Copyright © 1979 by the National Wildlife Federation. Reprinted from *National Wildlife Press Release*, October 1979.
10. A World of Silken Lines. Michael H. Robinson, *Smithsonian*, October 1987.
11. Daddy Longlegs. Richard Headstrom, Vermont Institute of Natural Science, October 1979.
12. Pond Giants. Linda Versage, Vermont Institute of Natural Science, May 27, 1985.
13. A Threat Outdoors. Lisa Halvorsen, University of Vermont Extension Service, July 31, 1987.
14. A Homespun Remedy for Fleas. Susan Peery. Reprinted with permission from *The Old Farmer's Almanac*, published in Dublin, New Hampshire.
15. Snow Fleas. Annette Gosnell, Vermont Institute of Natural Science, March 11, 1982.
16. The Life Cycle of a Moth. Ned Smith. Copyright © 1970 by the National Wildlife Federation. Reprinted from *National Wildlife* magazine, October/November 1970.
17. Why Do Woolly Bears Cross the Road? Margaret Barker, Vermont Institute of Natural Science, November 8, 1985.
18. Tiger Swallowtail Butterfly. Wayne Hanley, Massachusetts Audubon Society.
19. How to Plant a Butterfly Garden. Ted Williams. *Sanctuary*, April 1984, Massachusetts Audubon Society.

WATER & AQUATIC LIFE

1. Water Supply: Roman and Modern. Polly Bradley. *Back-Yard Frontier*, April 1979, Massachusetts Audubon Society.
2. Plumbing: A Look Back. Kenneth Mirvis, *The Old Farmer's Almanac*, 1989.
3. Small Wetlands. George Laycock, *Audubon*, July 1990.
4. The Estuary. Anita W. Brewer, *Sanctuary*, July/August 1984, Massachusetts Audubon Society.

5. Another World, the Coral Reef. Bonnie Ross, Vermont Institute of Natural Science News Release, November 16, 1987.
6. Tsunami: When the Sea Quakes. Michael J. Mooney, *Americas*, Vol. 42, No. 4, 1990.
7. Intertidal Creatures. Catherine Kiorpes-Elk, *Habitat: Journal of the Maine Audubon Society*, July 1987.
8. The Pond. We are grateful to Michael J. Caduto for permission to use this essay, which appears on page 173 of this book, and which is excerpted from *Pond and Brook: A Guide to Nature in Freshwater Environments* by Michael J. Caduto, University Press of New England, Hanover, New Hampshire. Copyright © 1990 by Michael J. Caduto.
9. Spring Peeping. Beth Ann Howard, Vermont Institute of Natural Science news release, April 14, 1986.
10. Wood Frogs. Beth Ann Howard, Vermont Institute of Natural Science news release, April 19, 1984.
11. Turtles. Excerpted from *The Year of the Turtle*, copyright © 1991 by David M. Carroll. Used by permission of Camden House Publishing, Charlotte, Vermont.
12. Outwitting Busy Beavers. Paul Fournier, *Maine Fish and Wildlife*, Winter 1989–90.
13. The Otter—Northwoods Playboy. Donald Wharton, *The Conservationist*, November/December 1989, magazine of New York State's Department of Environmental Conservation.
14. Soaking It All In. Jeri Galbraith. Copyright © 1983 by The National Wildlife Federation. Reprinted from *National Wildlife* magazine, April/May 1983.
15. Fish Sounds. William J. Kelly, editor, *Seaword*, Vol. 15, No. 4 (Fourth Quarter, 1987), Mystic Marinelife Aquarium, a division of Sea Research Foundation, Inc., Mystic, Connecticut.
16. Schooling. William J. Kelly, editor, *Seaword* (First Quarter, 1987), Mystic Marinelife Aquarium, a division of Sea Research Foundation, Inc., Mystic, Connecticut.
17. The Elegant Tunas. William J. Kelly, editor, *Seaword*, Vol. 12, No. 1 (1984), Mystic Marinelife Aquarium, a division of Sea Research Foundation, Inc., Mystic, Connecticut.
18. Landlocked Salmon. Vermont Fish and Game Department.
19. Fierce Fish. Copyright © 1981 by the National Wildlife Federation. Reprinted from *National Wildlife Press Release*, February 7, 1981.
20. Goldfish. *The Indoor Naturalist*, copyright © 1986 by Gale Lawrence. Published by Prentice Hall Press, a division of Simon and Schuster, New York, New York.
21. Killer Whale. Leonard J. Aube. Reprinted by permission from *Sea Frontiers* (November-December 1988), copyright © 1988 by the International Oceanographic Foundation, 4600 Rickenbacker Causeway, Virginia Key, Miami, Florida 33149.
22. Squid. Thomas A. Lewis. Copyright © 1988 by the National Wildlife

Federation. Reprinted from *National Wildlife* magazine, October/November 1988.

23. The Lobster. William J. Kelly, editor, *Seaword* (First Quarter, 1983), Mystic Marinelife Aquarium, a division of Sea Research Foundation, Inc., Mystic, Connecticut.

24. Lobstering in Winter. Raquel D. Boehmer and Peter J. Boehmer, *Seafood Soundings*, January/February 1990, Monhegan Island, Maine.

PLACES

1. Reindeer in Lapland. Will Curtis.
2. Iceland. Jane Curtis.
3. Harbor Watching. Jane Curtis.
4. The Right Animal in the Right Place. Jane Curtis.
5. The Isle of Jersey. Jane Curtis.
6. A Highland Road. Jane Curtis.
7. London with Grandsons. Jane Curtis.
8. Hadrian's Wall. Jane Curtis.
9. St. Patrick's Mountain. Jane Curtis.
10. Submarine Canyons. William J. Kelly, editor, *Seaword* (Third Quarter, 1988), Mystic Marinelife Aquarium, a division of Sea Research Foundation, Inc., Mystic, Connecticut.
11. Up in the Woods. Will Curtis.
12. Country of the Hill Folk. Jane Curtis.
13. A Cold Day in Yellowstone. Jane Curtis.
14. Fall Comes Early in Denali National Park. Jane Curtis.
15. Year-Round Alaska. Copyright © 1987 by Mary Shields. Reprinted from *Small Wonders: Year-Round Alaska*, published by Pyrola Publishing, Fairbanks, Alaska.
16. Patagonian Andes. Jane Curtis.
17. Limitless Argentine Pampa. Jane Curtis.
18. The Mysterious Mayans. Jane Curtis.
19. The Mayans Played a Tough Game of Basketball. Will Curtis.

HUMAN NATURE

1. Whiskey and Rum. Eleanor Ott, Vermont Folklife Center and Center for Research on Vermont, February 1985.
2. Cheese. Eleanor Ott, Vermont Folklife Center and Center for Research on Vermont, May 1986.
3. Sneezing. Eleanor Ott, Vermont Folklife Center and Center for Research on Vermont, June 1986.
4. Hiccups. Eleanor Ott, Vermont Folklife Center and Center for Research on Vermont, December 1984.
5. Stress Management. Monica Porter, University of Vermont Extension Service, 1981.
6. Joe Ranger. Jane Curtis.
7. Rainy-Day Hiker. Barney Crosier, *Rutland Herald/Times Argus*, March 1981.

8. End of the Tether. John H. Mitchell, *Sanctuary*, November 1987, Massachusetts Audubon Society.
9. Recollections of a Farm Woman. Reprinted from *The Last Yankees: Folkways in Eastern Vermont and the Border Country*, by Scott E. Hastings, Jr., copyright © 1990 by the Estate of Scott E. Hastings, Jr., by permission of University Press of New England.
10. Clog Almanacs. Eleanor Ott, Vermont Folklife Center and Center for Research on Vermont, December 1986.
11. Dwellings. *The Indoor Naturalist*, copyright © 1986 by Gale Lawrence. Published by Prentice Hall Press, a division of Simon & Schuster, New York, New York.
12. Bless You, Old Barn Door. Barney Crosier, *Rutland Herald/Times Argus*, 1980.

PLANETS & SPACE

1. Keeping Tabs on a Wandering Venus. Gale Lawrence, *Rutland Herald/Times Argus*, February 1986.
2. Contemplating Constellations. Marie Levesque Caduto, Vermont Institute of Natural Science, January 25, 1991.
3. Meteors. Jane Curtis.
4. Understanding the Aurora. Dr. William Stringer and Linda Schreurs, Geophysical Institute, University of Alaska.
5. Clouds. Eileen Docekal, *Sierra* magazine, April 1985.
6. Ever Wonder Why There's Wind? Gale Lawrence, *Rutland Herald/Times Argus*, March 1985.
7. Earthquakes—Adjusting to Stresses. Gale Lawrence, *Rutland Herald/Times Argus*, November 13, 1983.
8. Summer Clouds. Jenepher R. Lingelbach, Vermont Institute of Natural Science, June 28, 1985.
9. Weather Lore. Will Curtis.

THE ENVIRONMENT & CONSERVATION

1. Gaia: The Earth. Eleanor Ott, Vermont Folklife Center and Center for Research on Vermont, January 1988.
2. The Manager of the Biosphere. Jay D. Hair. Copyright © 1984 by the National Wildlife Federation. Reprinted from *International Wildlife* magazine, November/December 1984.
3. The End of Living and the Beginning of Survival. Chief Seattle.
4. The Tropical Equation. T. H. Watkins, *Wilderness*, Winter 1989, The Wilderness Society, Washington, D.C.
5. The Unpayable Costs of Global Warming. Donella Meadows, *Valley News*, October 10, 1989.
6. Vermont Forests Weather the Acid Test. Experimental Station Acid Rain Laboratory, University of Vermont.
7. George Perkins Marsh. Jane and Will Curtis, and Frank Lieberman, *George Perkins Marsh*, published by Countryman Press. Copyright © 1982 by The Woodstock Foundation.

8. We Need Nothing Short of a Revolution. Donella Meadows, *Valley News*, June 1990.

9. It Will Take Time to Perfect Recycling. Donella Meadows, "The Global Citizen," *Valley News*, November 13, 1990.

10. Don't Take It Anymore. Don Hickman, *Sanctuary*, January 1989. Massachusetts Audubon Society.

11. Household Waste Disposal. Don Hickman, *Sanctuary*, January 1989, Massachusetts Audubon Society.

12. Indoor Air Quality. William Turner and Terry Brennan, *Habitat: Journal of the Maine Audubon Society*.

13. Sick Building Syndrome. Linda Lee Danidoff. Copyright © 1989 by *The Amicus Journal*, Winter 1989, a publication of the Natural Resources Defense Council. Reprinted with permission.

14. Dry Homes. Bill Christiansen, *News from County Agents*, February 17, 1989, University of Vermont Extension Service.

15. Wood Smoke. "Home Safe Home," Natural Resources Council of Maine.

16. Where There's Fire, There's Smoke. Thomas A. Lewis. Copyright © 1988 by the National Wildlife Federation. Reprinted from *National Wildlife* magazine, October/November 1988.

17. Our Fate Depends. . . . Donella Meadows, "The Global Citizen," *Valley News*, March 11, 1989.

18. In the End. Editorial, *Valley News*, January 29, 1968. Originally written by Kenneth Ross, Austin, Minnesota.